About Island Press

Island Press, a nonprofit organization, publishes, markets and distributes the most advanced thinking on the conservation of our natural resources—books about soil, land, water, forests, wildlife and hazardous and toxic wastes. These books are practical tools used by public officials, business and industry leaders, natural resource managers and concerned citizens working to solve both local and global resource problems.

Founded in 1978, Island Press reorganized in 1984 to meet the increasing demand for substantive books on all resource-related issues. Island Press publishes and distributes under its own imprint and offers these services to other nonprofit organizations.

Support for Island Press is provided by The Geraldine R. Dodge Foundation, The Energy Foundation, The Charles Engelhard Foundation, The Ford Foundation, Glen Eagles Foundation, The George Gund Foundation, William and Flora Hewlett Foundation, The James Irvine Foundation, The John D. and Catherine T. MacArthur Foundation, The Andrew W. Mellon Foundation, The Joyce Mertz-Gilmore Foundation, The New-Land Foundation, The Pew Charitable Trusts, The Rockefeller Brothers Fund, The Tides Foundation and individual donors.

Saving All the Parts

Saving All the Parts

Reconciling
Economics
and the
Endangered
Species Act

Rocky Barker

ISLAND PRESS
Washington, D.C.
Covelo, California

Copyright © 1993 by Rocky Barker

All rights reserved under International and Pan-American
Copyright Conventions. No part of this book may be repro-
duced in any form or by any means without permission
in writing from the publisher: Island Press, Suite 300,
1718 Connecticut Avenue, N.W., Washington, DC 20009.

ISLAND PRESS is a trademark of The Center for Resource Economics.

The author is grateful for permission to include portions of this work that
had their origins in news stories written by the author for the *Idaho Falls
Post Register* from 1985 to 1992. All figures were drawn by Jerry Painter.
All chapter-opening illustrations were drawn by R. D. Dye.

Library of Congress Cataloging-in-Publication Data

Barker, Rocky.
Saving all the parts : reconciling economics and the Endangered
Species Act / Rocky Barker.
p.   cm.
Includes bibliographical references (p.   ) and index.
ISBN 1-55963-202-X : —ISBN 1-55963-201-1 (pbk.)
1. Biological diversity conservation—Law and legislation—
Economic aspects—Northwest, Pacific.   2. Biological diversity
conservation—Law and legislation—Economic aspects—Rocky
Mountains.   3. Economic development—Environmental aspects.
4. United States—Economic policy.   I. Title.   II. Title:
Endangered Species Act.
QH76.5.N97B37   1993
333.95'11—dc20   93-4640
                                                         CIP

Printed on recycled, acid-free paper

Manufactured in the United States of America

10   9   8   7   6   5   4   3   2   1

This book is dedicated to Jeff Wilson

Contents

Preface

THIS book examines the ecological and economic forces at work in the Pacific Northwest and northern Rockies of the United States that will shape biodiversity protection in this country and perhaps the world. It also analyzes the trends in natural-resource management, land-use planning and economic development that can lead to a future in which the region, the nation and the world can sustain economic activity without the loss of the natural values that make them special.

The book grew out of a six-month project undertaken by the author and staff of the *Idaho Falls Post Register*. The project, which culminated in a seven-part series running from December 23 through December 31, 1990, examined the effects of the Endangered Species Act on the people of the Pacific Northwest, concentrating on its effects on eastern Idaho, where the newspaper is located.

Some readers may wonder why there is not more detail about the ancient forest controversy and the coastal forest debates of Washington, Oregon and northern California. There already are many good books on those subjects, and I covered the issues only as far as I thought necessary to fit them into the larger context. Also, my view of the Pacific Northwest comes from its eastern boundaries, and with that comes the inevitable bias toward issues that affect the entire Columbia River watershed, such as salmon management, and those issues in my backyard, such as Greater Yellowstone Ecosystem protection.

Since the newspaper project was a collaborative effort, I am indebted to the reporters who worked on that project with me. Their research contributed to the strength of the series and filtered through to the book. They include Kevin Richert, Stuart Englert, Margaret Wimborne, Molly O'Leary Cecil, Rob Thornberry, Pat Durkin, Jon Jensen and Candace Burns. Several editors at the *Post Register* also helped shape the endangered species series; these include Roger Plothow, Ken Retallic, Genie Arcano, Kristie Jones and Janet Donnelly. Pat Ford, who was a contributing editor for *High Country News*, also helped me in portions of the chapters on salmon.

The idea for the series originated with Jay Gore, now a U.S. Forest Service biologist in Washington, D.C. Associated Press correspondent Tad Bartimus encouraged me to write the book and helped in ways I can never repay. Hadley Roberts also urged me to turn the series into a book. Tim Palmer introduced me to Island Press and Barbara Dean, and I am very grateful to both.

I owe a special debt of gratitude to Jerry Brady, publisher of the *Post Register*, who demonstrated a remarkable commitment to environmental journalism when he approved and underwrote the endangered species series. His continued encouragement made it possible for me to write the book while continuing my reporting duties. He also deserves my thanks for reading the manuscript, as do Ken Retallic and Tim Crosby.

Many people in the environmental movement, state and federal agencies, special-interest groups and industry also went out of their way to help me during my research. (All direct quotes in the book are from my interviews with the people quoted, unless otherwise noted or sourced in the bibliography.) Many are quoted as sources in the book, but several deserve special mention, including Sharon Blair and Rick Atami of the Bonneville Power Administration, Merritt Tuttle of the National Marine Fisheries Service, Bob Ekey of the Greater Yellowstone Coalition, Ed Chaney of the Northwest Resource Information Center, Jack Trueblood of the Idaho Department of Fish and Game, Joan Anzelmo and Marsha Karle of Yellowstone National Park, Jim Riley of the Intermountain Forest Industry Association, Mike Tracy of the Idaho Farm Bureau and Dan Funsch of the Alliance for the Wild Rockies.

My wife, Tina, and my three children sacrificed for a year as I worked on the book, making it possible for me to do so without leaving work. Barbara Youngblood, my developmental editor at Island Press, led me through the process of writing a book with patience, diplomacy and skill I will never forget.

Finally, I must thank Rob Brady, retired publisher of the *Post Register*, for bringing me to Idaho and nurturing my love for this special place on the earth. His optimistic vision of a future in which humans and other species live harmoniously in this sometimes harsh land serves as my guiding beacon into the twenty-first century.

Saving All the Parts

1. Endangered Species, Embroiled Region

TUCKED in the sundown shadows of the Sawtooth Mountains in central Idaho, Jack See has built a life many dream about. As owner of the historic Redfish Lake Lodge near Stanley, Idaho, See has prospered by serving trout dinners, renting boats, shuttling hikers and providing lodging for visitors to the Sawtooth National Recreation Area. The short, blond, forty-something lodge owner with a distinctive Fu Manchu mustache and his wife, Patty, have made Redfish Lake Lodge one of Idaho's best-known attractions. Idaho has grown steadily during the more than twenty years since See and his father took over the Redfish Lake Lodge, and the resort has benefited.

Nearby Boise has become one of the fastest-growing cities in the West, attracting several computer-technology manufacturing plants and serving as corporate headquarters for companies such as Boise Cascade, Albertson's Supermarkets and Micron Technologies. Ketchum and Sun Valley to the south have become the conduit for a stream of former Californians who have spilled into the state so they can have places like Redfish out their back doors. On summer weekends Redfish Lake Lodge is filled mostly with Idahoans heading for backcountry jaunts into the nearby White Cloud Mountains, sailing on the lake or simply sitting on the porch in front of the lodge to sun themselves and admire towering Mount Heyburn. As the main jumping-off point into the Sawtooth Wilderness, the lodge has become almost a headquarters for the state's environmental community. Each spring, a week before See opens the lodge to the public, the Idaho Conservation League holds its annual "Wild Idaho" conference there. In August the group returns

4

to hold a vigil at nearby Redfish Lake Creek, awaiting the return of the few remaining sockeye salmon. Campgrounds encircle the crystal-clear lake and the campers buy their groceries in See's small store and gas for their boats at his marina.

The placemats in the lodge explain how Redfish Lake got its name from the sockeye salmon and kokanee salmon that spawn in the lake. The fish turn bright red when they spawn, and historically the entire shoreline would blaze with color as thousands of fish filled the shallows to spawn. The kokanee live year-round in Redfish but the sockeye have always been only temporary residents. Young sockeye stay in the lake a little more than a year after hatching on its windswept shoals before migrating down the Salmon River to the ocean. If they are lucky, they will return in another two years. Sockeye were one of the attractions at Redfish Lake Lodge. In the fall, when the fish spawned, See used to take tourists on pontoon boat tours of the spawning beds. He would turn off the engines and he and his guests would marvel at the spawning pairs acting out their annual ritual. In the early 1980s, See realized that something was wrong. Suddenly the sockeye were gone. Civilization had caught up with the Redfish Lake sockeye. The eighth in a series of dams on the Snake and Columbia rivers was completed in 1975 and biologists said it was the final straw for Redfish's sockeye. Fewer and fewer fish were able to make the long trip between the ocean and the lake and back with the added perils posed by the dams. From 1975 through the early 1990s, the number of sockeye sharply declined; in 1989 none made the trip.

In 1991, the National Marine Fisheries Service (NMFS) listed Snake River sockeye salmon as an endangered species under the federal Endangered Species Act. Since Redfish Lake is the only spawning ground left for sockeye in the Snake River drainage, the NMFS may take sweeping actions to protect the fish. It may limit the boating that attracts many of the campers to Redfish. Kayaks and canoes could replace powerboats and water-skiers. Idaho's Department of Fish and Game may discontinue stocking rainbow trout that provide sport for fishermen because the trout compete with young sockeye. Many of the recreational activities that attract people to Redfish Lake Lodge could eventually cease. Obviously, this will directly hurt See's business.

It might not have to happen, See hopes. Perhaps the fish can be saved without severe constraints on lake use. After all, it wasn't at Redfish Lake that the sockeye became endangered, See says; it's the rest of its long and precarious habitat that has become lethal. See may have to make only minor adjustments in his lifestyle and his business to aid the effort to save sockeye, and he says he's ready to do that. The sockeye are an important part of his life and worth some sacrifices; his lifestyle is tied to the raw, wild character that the sockeye represents. Though he may be more tied to this than others, many of the residents and economies of the Pacific Northwest are caught in the same dilemma as See. They want to preserve aspects of the Pacific Northwest that make it special and still earn a living that allows them to stay in the region.

In the 1990s, efforts to balance human progress with preservation of the earth and its wild inhabitants will affect virtually every resident of the Pacific Northwest. How the Endangered Species Act is enforced in this decade could affect everything from electricity rates and home-building costs to how much water is available to irrigate Idaho's famous potatoes.

Called "an American hinterland" by University of Idaho historian Carlos Schwantes in his wonderful history *The Pacific Northwest,* the region stretches from the vast old-growth forests of coastal Oregon and Washington to the Continental Divide country of Montana, to the blooming deserts of eastern Idaho and the natural wonders of Yellowstone. The natural character of most of the continental United States is already gone, replaced by megatropolises, giant farms and sprawling suburbs. Humans have spread throughout the United States, changing the landscape from one shaped by nature to areas mostly fashioned by human development. A few pockets remain, such as the Everglades and the Adirondacks, but even in the sparsely populated regions of New England, the upper Midwest and the South, massive timber harvests of the past or early farm development transformed the natural ecosystems into different, less diverse systems. Much of the interior West remains intact only because its lack of water and resources makes it useless to all but a few ranchers, miners and oil developers. Even so, the natural character of the Great Plains and the deserts has been changed by the replacement of buffalo with cattle and by past over-grazing, which has altered plant communities.

In the Pacific Northwest, human development was mostly concentrated along waterways and in the vicinity of ports for nearly 200 years. The mountainous topography near places such as Puget Sound in Washington and the Willamette Valley in Oregon acted as barriers to urban growth and gave even urban residents a sense of

6

living next to the wilderness. Because of its isolation, it was not until the 1980s that the Pacific Northwest reached the stage in its development where wild, species-rich landscapes and waterways were changed forever by bulldozers, hydroelectric dams and condominiums. Even today, much of the hinterland remains.

The 1973 Endangered Species Act, which may shape the future of the Pacific Northwest, is designed to prevent the extinction of species, both in the United States and worldwide. Since Europeans arrived in the Americas, 500 plant and animal species have become extinct in what is now the United States, including the passenger pigeon, the great auk and the California grizzly bear. The rate of extinctions in the nation and the world is increasing, and with the loss of each species, humans lose priceless information and genetic material. Today's extinct species may have been tomorrow's wonder drug or miracle food source.

As of April 1, 1992, some 277 domestic animals and 243 domestic plants were protected under the act as endangered. Another 97 animals and 64 plants were classified as threatened. The Endangered Species Act also recognizes 528 species in other nations as endangered or threatened and thus protected under the act.

The law requires the U.S. Fish and Wildlife Service (FWS) and the NMFS to list species and subspecies endangered of extinction or threatened to be endangered of extinction. The two agencies must attempt to restore viable populations of the listed species by protecting the remaining individuals and their habitat.

The future of the Endangered Species Act may rest on the outcome of the Pacific Northwest's environmental battles. In particular, the future of Pacific salmon and the act may be inseparable. With several stocks of Snake River salmon already placed on the endangered species list, restrictions imposed by the act literally could cover all activities within the Columbia River watershed. After the NMFS listed Snake River sockeye salmon as an endangered species, it listed Snake River chinook salmon stocks as threatened. Suddenly, because of the potential changes in operations of hydroelectric dams that provide much of the region's electricity, a law that had affected only the region's rural residents was extending its reach to impact the lives of every person who turned a light switch.

The effects of listing salmon won't be contained west of the Rockies and north of Big Sur. The fallout from the decision will spread beyond the borders of the Pacific Northwest to the entire nation. What happens to species such as the spotted owl, grizzly bear and Pacific salmon may indicate how serious Americans are about preserving their natural heritage. And as more and more people are asked to change their lives to make room for endangered species, the powers of the Endangered Species Act will be tested.

7

In the Pacific Northwest, the ramifications are widespread. Most plans to restore salmon stocks would force the operators of the Columbia and Snake river hydroelectric dams to alter their operations during spring salmon migration, costing Northwest ratepayers an additional 2 percent to 30 percent on their electric bills over the next decade. The region's $5 billion irrigated farming industry would have to extend pumping pipes and place fish screens on head gates and may have less water to use. Fishermen, already facing strict limits, may be forced to leave their boats in harbor or their nets on riverbanks for years to come. Navigation on the Columbia and Snake rivers may be restricted for at least a portion of the spring. Logging, grazing and mining on public lands could be curtailed.

Each region of the United States has a defining characteristic, says historian Schwantes—for example, in the Southeast, the history of the pre–Civil War plantation culture ties the region together. In the Pacific Northwest, Schwantes says, the environment is a defining feature—the rugged mountains, deep river canyons and wide, desolate, high-elevation deserts. The appreciation and enjoyment of the region's spectacular setting unite the region's people as nothing else does. Even as politics and philosophy divide them, their sense of place binds them together.

Although places such as Puget Sound have become heavily populated, manipulated and tamed, the powerful beauty of the region's natural wonders dominates both the emotional and physical landscape. Mount Rainier rises above the Seattle skyline calling urban hikers to climb to its peak. Portland fishermen still catch salmon in the Willamette and Columbia rivers. After work, Boise mountain bikers climb the mountains east of Idaho's fastest-growing city. On weekends, the region's urban residents head for the hills, rivers and lakes to play.

The giant tracts of still-uninhabited roadless land are important to the lifestyles of many Pacific Northwest residents who hike, climb, ski, ride horseback, hunt, fish, snowmobile, mountain bike or motorbike in the backcountry. A 1989 study by University of Idaho geography professors Harley Johansen and Gundars Rudzitis showed that counties containing federally designated wilderness areas outpaced other rural western counties in growth during the last thirty years. The study showed that most people said they moved to the wilderness county because of the enhanced quality of life. Rudzitis, in a draft paper written in 1992, said rural counties of the West in the 1980s grew faster than the rural counties of all other regions of the country and that the West's wilderness counties grew even faster than other rural counties—a remarkable 24 percent, or six times the national average of 4 percent for nonmetropolitan counties. Even for people who never set foot off the highway, the value of undeveloped mountain ranges filled with grizzly bears, caribou, bighorn sheep and elk plays an important role in their own ties to the region. The thought of wild animals such as these triggers striking recollections for people, whether based on an encounter in a national park or something seen on a television nature program.

During the 1980s the Pacific Northwest was one of the fastest-growing regions in the country. Idaho, Washington and Oregon were the third, fourth and fifth fastest-growing states in the 1980s, behind only neighboring Nevada and Alaska. More than 8 million people now live in the three states and western Montana. The growth rate appears to be accelerating rather than slowing into the 1990s and adds to the challenge of preserving the environment and meeting the needs of a growing economy.

Culturally, the biggest challenge comes as a consequence of the long-term shift from a rural land to an urban one. The population is concentrated in urban areas, leaving most of the region still relatively unoccupied. In his most recent book, *In Mountain Shadows,* Schwantes notes about Idaho that of its 53 million acres only one-third of 1 percent are urban, where most of its population growth is taking place. "The juxtaposition of the modern metropolis and the hinterland is, in fact, the defining quality of life in modern Idaho," Schwantes writes. The same could be said for the entire region.

The Pacific Northwest region can be defined in many ways. The Bonneville Power Administration (BPA), which markets cheap, mostly federal hydroelectric electric power to the Pacific Northwest, defines the region as its service area, about 300,000 square miles in Oregon, Washington, Idaho, Montana, Nevada, Utah and Wyoming. Since its power grid has contributed to the regional nature of the area's economic development, this is a fair definition. The Columbia River drainage, perhaps a more natural regional boundary, covers 259,000 square miles in the seven states and British Columbia, an area larger than France and nearly as large as Texas. Conservation biologists, taking an ecosystem approach to the region, might define it differently, adding in areas where 9 grizzly bears live east of the Continental Divide. Other huge, largely intact ecosystems, such as the Northern Continental Divide Ecosystem beginning in Alberta and running into Montana, and the Greater Yellowstone Ecosystem, in Montana, Idaho and Wyoming, straddle the divide and naturally are tied to the Columbia drainage. Economically and even culturally, many people include Alaska in the region.

The climate of the Northwest is a study in contrasts. Along its coastline and east along the Canadian border, heavy rainfall sustains rain forests—the most productive in the world. But in the southeastern section, rainfall of only seven inches annually is not uncommon. The desert sections of Idaho, Oregon and Washington are more like the West's interior than what is traditionally identified with the Pacific Northwest. Yet some of the most productive farmland in the West lies in these areas, sustained by irrigation from the snowmelt of the high peaks of the Rocky Mountains delivered by the mighty Columbia and Snake rivers.

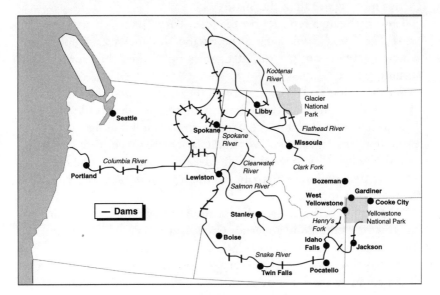

The Columbia River watershed reaches into Wyoming, Montana, Alberta and British Columbia.

No matter how you look at the Pacific Northwest, the Columbia River system is the thread that binds the region together. The Columbia is the beneficiary of 150 tributaries large enough to be called rivers. In an average year the Columbia dumps enough water into the Pacific Ocean to cover 196 billion acres of land with one foot of water. The Columbia River and its tributaries contain 30 percent of the hydro-electric power production potential on the North American continent. The Columbia and Snake rivers deliver more goods by water than any other waterway on the continent save for the Mississippi. The Columbia itself is the fourth-largest river in North America. Only the Mississippi, St. Lawrence and Mackenzie rivers are larger. Yet even they can't match the Columbia's electricity-generation capacity because they don't drop in altitude so quickly.

The river begins in British Columbia eighty miles north of the international boundary in Columbia Lake. It runs 1240 miles through forest, desert, coulee and mountains before reaching the Pacific Ocean. Its largest tributary, the Snake River begins in Yellowstone National Park and runs 1038 miles through Wyoming, Idaho and Washington, where it meets the Columbia. The Snake alone drains an area of 109,000 square miles. In the western United States, only the Columbia carries more water than the Snake.

Timothy Egan, a reporter for the *New York Times* and author of the book *The Good Rain,* defines the Pacific Northwest as simply "anywhere a salmon can get to." If Egan is right, then the Northwest is shrinking fast.

For the Indians who first inhabited the region, salmon were a source of sustenance and stability. Each spring native peoples would prepare a ceremonial feast to thank the Creator for this miraculous blessing. In 1805, in northern Idaho, Lewis and Clark were saved from starvation by the Nez Perce, who fed them dried salmon. Less than two centuries after the two explorers first entered the region, the salmon that saved them are disappearing.

"More than perhaps any other symbol, the salmon distinguishes the Northwest," said Idaho governor Cecil Andrus in a speech to the 1991 salmon summit held in Boise, Idaho. "Some of the fundamentals of life in the Northwest, some of the very reasons many of us live in this part of the country, are dying or threatened along with our salmon."

For more than 10,000 years, since the last Ice Age, salmon have undertaken an annual odyssey even more amazing than Homer's story of Ulysses. Salmon have been leaving the Pacific Ocean in the spring to migrate up thousands of streams to spawn throughout the Pacific Northwest, north to Alaska and across the ocean in Russia. Their strong and still not completely understood homing instinct has drawn them back to the tributaries where they hatched. Small numbers have strayed to other streams, filling every available river in the region with salmon as far inland as the fish could swim without insurmountable barriers. Historically, they spawned as far east as Montana and perhaps Wyoming, though most scientists believe Shoshone Falls on the Snake River near Twin Falls, Idaho, was the final eastern barrier. Even today, chinook and sockeye salmon still swim 930 miles and climb 6500 feet in altitude to spawn in the headwaters of the Salmon River in the Sawtooth Valley. Every spring, these remarkable fish swim up the Columbia River through four dams and then turn right up the Snake River and negotiate another four dams before entering Idaho. Then, drawn by the slightest scent of their home water, they turn left to swim up the Salmon River. About sixty days after they enter the Columbia River, they reach Stanley, Idaho. The chinook, if they're wild, spawn in the headwaters of the Salmon River or in tributaries such as Alturus Lake Creek, 930 miles from the ocean. Hatchery stocks are captured and artificially spawned in the new Sawtooth National Hatchery. The sockeyes, or what is left of them, head for Redfish Lake. But today, few if any sockeye are actually allowed to spawn naturally. Instead they are captured and artificially spawned. Scientists believe that their numbers are so low and their natural habitat, especially their migration route, so lethal that they are better off in hatcheries.

Once the eggs hatch in stream or lake, the young salmon fend for themselves in the immediate area, often for more than a year. Then when the time is right, from early to late spring for the Salmon River salmon, they begin their migration back to the ocean. When white settlers arrived in the Pacific Northwest in the 1850s, 8 million to 16 million salmon ran up the Columbia, and the runs filled the rivers continuously from March to December. Today, only spring, summer and fall chinook runs survive, and less than 2.5 million return.

When the young fish are ready to migrate to the ocean, their bladders enlarge and their body shape alters. This smolting process prepares them for life in salt water. Traveling mostly at night, driven by the current, the smolts are flushed downstream by the runoff from the melting snow. Historically, the trip through the Snake and Columbia rivers took less than five days. Today the reservoirs behind the dams that now catch the spring runoff slow the trip down to seventeen to thirty days. Most of the salmon die in Lower Granite Reservoir, on the Washington-Idaho border. There they either get lost in the forty-mile pool of slack water, are eaten by predators, die of bacterial kidney disease or get shredded in the turbines of the hydroelectric plant. Most of those that survive are caught and placed in barges, which then take them past the dams. They are dumped in the Columbia Estuary below Bonneville Dam near Portland, Oregon, and many just disappear. This manmade system has altered the smolts' ability to make the transition to salt water and has dramatically cut the number that finally reach the sea.

Once there, the salmon swim thousands of miles for one to three years, scattering all the way to the Gulf of Alaska and Asia. They begin feeding on plankton and then progress to shrimp, herring and anchovies. When they reach maturity they head back to the coastal waters of the Pacific Northwest for the return to their home rivers. Many swim more than 10,000 miles during their short life cycle.

"The salmon runs are a visible symbol of life, death and regeneration, plain for all to see and share . . . ," wrote Canadian nature writer Roderick Haig-Brown. "If there ever was a time when the salmon no longer return, man will know he has failed again and move one step nearer to his own final disappearance."

The controversy concerning salmon restoration and its potential effects on the Pacific Northwest economy only amplifies the debate about the Endangered Species Act that started with the northern spotted owl. The case of the spotted owl has become the classic endangered species confrontation. The northern spotted owl is a medium-size brown owl with dark eyes and white spots on its head and neck and white mottling on its underside. Like most owls it is nocturnal and feeds on forest-dwelling rodents. It is one of three subspecies of the spotted owl; the other two are the California spotted owl and the Mexican spotted owl.

The northern spotted owl, a skillful but specialized predator, lives in old-growth forests and mixed stands of mature and old-growth trees usually 200 years old or older. The old-growth forests of the Pacific Northwest, where the owls live, have a multilayered canopy of trees, large-diameter live trees, some with broken tops, and both dead and dying trees standing and fallen on the ground. These unique charac-

teristics create an ecosystem that supports a wide variety of animals. Fishers, pine martens, Olympic salamanders, marbled murrelets and many other species depend on old-growth stands for survival. Scien-   13
tists now argue that salmon stocks also depend on old-growth watersheds for their survival in many rivers. When humans began settling the Pacific Northwest 150 years ago, much of the coastal area was covered by thick, old-growth stands of Douglas fir, but today only about 10 to 20 percent of that old-growth forest remains.

Much more remained in the late 1960s when scientists at Oregon State University began studying northern spotted owls. When Professor Howard Wight and his graduate student Eric Forsman started surveying the northern spotted owl's territory, they quickly established its need for old-growth forest for survival. Wight and Forsman also determined in 1972 that logging of the ancient forests was threatening the bird with extinction. Nearly everywhere that they found owl nests, there were timber sales pending to cut down the trees. When the Endangered Species Act was passed in 1973, the northern spotted owl was listed as a candidate for threatened or endangered species protection, based primarily on Forsman's and Wight's research. At the time, the two scientists recommended protecting 300 acres of old-growth timber around each pair of owls, a recommendation that was ignored by the U.S. Forest Service (USFS) and the Bureau of Land Management (BLM).

The Oregon Wildlife Commission listed the owl as threatened in 1975, but the listing had little effect on logging. In 1981, the U.S. Fish and Wildlife Service (FWS) decided against listing the owl as threatened at a time when the harvest level on federal lands was being accelerated. Finally, in 1990, the owl was listed. The federal government finally acknowledged that the future of the owl and the future of old-growth forests were one and the same.

Dozens of rural communities now face the end of the massive harvesting of old-growth forests that peaked in the late 1970s and continued above the level even many foresters said was sustainable into the late 1980s. The northern spotted owl has taken most of the blame for halting the harvest, but even the most optimistic foresters in the region said such harvests could not continue forever. The Douglas fir forests of Oregon, Washington and northern California could withstand only about ten more years of harvesting at the levels of the 1980s before the trees that date back to the time of Christ would be exhausted. Then

14

loggers would have to move on to the second-growth forests that were planted from the 1940s through 1960s. Timber harvests have been so high for so long that no matter what happens to the ancient forests, loggers and the rural communities they support face a long wait for the next generation of trees to mature. These trees are not expected to be ready to cut until well into the next century, leaving a large gap in the supply of timber for jobs and homes.

In the forests of Idaho and Montana, the story is the same, only owls are replaced as the focus of anger and frustration for the timber industry and loggers by grizzly bears, caribou and wolves. All three of these large mammals depend on big, wide-open country with huge stands of timber for cover. To protect them, some of the timber that loggers would otherwise cut must be left standing. Once again, the options for protecting both wildlife habitat and jobs have been limited by a massive clearcutting campaign. Heavy timber cutting, primarily on private land in the 1980s, has pockmarked the mountainsides of Montana and Idaho, filling streams with muddy sediment and pressing civilization into the heart of the region's remaining wilderness. Even though nowhere near the volume of timber cut on the Cascades was harvested in the Rockies, the nature of the land and climate amplified the effects. The forests of the northern Rockies are significantly drier than the lush coastal stands of the Cascades and the growing season is shorter. Even though the timber industry began harvesting trees in the northern Rockies later than in the Cascades, the effects of the cutting will likely last longer because the trees will take so long to grow back.

John Osborn is a Spokane doctor who, when he's not taking care of AIDS patients, is battling the Forest Service and timber industry's efforts to cut over wild lands in Idaho, Washington and Montana. He compares the cutting of the last forty years to the elimination of the great forests of the North Woods of Wisconsin and Minnesota by lumber barons a century ago. In fact, some of the same companies are involved. "We are at the end of the timber frontier," Osborn says.

The timber frontier began on the East Coast and by the middle of the nineteenth century had moved to the great forests of the upper Midwest in Michigan, Wisconsin and Minnesota. In less than fifty years all but a few small groves of the great North Woods forest had been chopped down, floated to mills and shipped to markets across America. Once they had leveled the North Woods, the logging companies moved on to the Pacific Northwest.

These lumbermen were not evil-hearted villains stealing the future. They were meeting the needs of an expanding America, cutting wood to build homes and cities. Placed in the context of their times they were giants, helping to make the United States a world economic power. In 1858, Frederick Weyerhaeuser took a debt-ridden lumber company in Rock Island, Illinois, and turned it into a timber empire that eventually stretched across the continent. He cut thousands of acres of white pine in Minnesota and Wisconsin through much of the late 1800s, picking up land holdings and partners as the forests of the region were quickly being turned into lumber and houses. In 1900, at the age of sixty-five, he took his greatest gamble and with fifteen partners purchased 900,000 acres of virgin timber in the Pacific Northwest from the Northern Pacific Railroad for six dollars an acre. Weyerhaeuser's grandson Phillip saw what his family and other timber men had done to the North Woods and sought a different course for the company in the Pacific Northwest. In 1941, Weyerhaeuser started a tree farm program to replace the trees the company cut. Today Weyerhaeuser has 5.6 million acres in tree farms and is the largest private landowner in the Pacific Northwest.

After World War II, the harvest of the Pacific Northwest's forests increased as a new housing boom required more and more wood. Unlike the cut-and-run tactics employed by Weyerhaeuser and other lumbermen in the Midwest, this new generation of timber executives and professional foresters set as a goal to cut no more wood than the forest grew annually. This sustained-yield concept, developed by Germans and brought to the United States by Bernard Fernow and Gifford Pinchot, was the basis of the science of forestry. The concept called for cutting down old, "decadent" natural forests that grew slowly and replacing them with human-planted, carefully propagated forests that grew faster and were easier to manage. These foresters saw recreation and wildlife habitat as desirable offshoots of their work, but only secondary to the scientific and economic task of growing trees for homes. From the beginning, many timber executives looked upon sustained-yield forestry as a goal rather than a limit. Economically, sustained-yield forestry always was a marginal endeavor that forced corporations to take a long-term view of their land and timber resources.

Now, a century later, the combined actions of the early barons and the scientific foresters who followed them have left little of the ancient

16

forests of the Northwest. John Osborn says despite the goal of cutting no more wood fiber than it can grow, the timber industry has operated since the 1800s as if there would always be more virgin forests to cut over the next ridge line. "What's over the next ridge line now is the Pacific Ocean," says Osborn.

Timber harvest on federal lands in Oregon and Washington hit its peak from 1987 to 1989, when more than 20 billion board feet of timber was cut. (It takes about 10,000 board feet to build a house.) The Forest Service, at the direction of the Reagan administration and Congress, dramatically increased the amount of timber it sold. The economic boom of the Reagan years had energized the lumber market worldwide, and the timber industry was cutting and selling every stick it could buy. Moreover, the pressure of leveraged buy-outs on Wall Street forced many timber companies to cut down as many trees as they could to prevent a corporate raider from buying stock of the company with junk bonds and then paying the bonds off by cutting and selling the trees themselves.

Much of the timber cut from private and state lands went to Japan, sometimes to subsidiaries of the lumber companies that were cutting the wood. Weyerhaeuser, for example, was shutting down mills in Oregon in 1990, despite timber supplies in a nearby tree farm, because it "can get three times the price selling logs for export than it can get domestically," an unnamed writer said in the spring 1990 edition of the *Inner Voice,* a publication of the American Forest Service Employees for Environmental Ethics. In 1989, Weyerhaeuser sold more than $250 million worth of raw logs to Japan, according to the Council on Economic Priorities. In addition, throughout the 1980s, timber industry jobs were disappearing in the region due to mill efficiency programs and modernization. Timber industry employment on the west side of the Cascades in Oregon and Washington dropped from 101,875 in 1980 to 88,000 in 1988, when timber harvest was the highest. Similar trends were evident in Idaho and Montana.

The Pacific Northwest went through a tremendous period of growth in the 1980s. Companies such as Microsoft, Boeing and other high-technology companies in all four states benefited from the Reagan boom, creating thousands of new jobs. Small businesses also grew, adding most of the new jobs that kept employment high as the population grew. Meanwhile, in the traditional extractive industries—wood products, mining and agriculture—employment was stagnant or declining. In Idaho, the timber and mining industries lost 7300 jobs from 1979 to 1989. During the same period, electrical and machinery manufacturing and service and health industry jobs increased by 30,600, according to the Idaho Department of Employment. Similar trends were found in the other Pacific Northwest states. Most of these new jobs were created by people starting their own businesses in the service industry or by marketing new products. Tourism grew in the region, and businesses such as Jack See's Redfish Lake Lodge expanded and new resorts were built. In each sector where new growth was realized, the attractions of the communities in the Pacific Northwest—clean air, open spaces, low crime—were major factors.

17

An increasing number of economists are arguing that preserving the region's quality of life is necessary to continuing its economic upturn. Whether providing tourism jobs or attracting people and businesses looking for a good quality of life, protecting the scenic amenities of the region are key to its long-term economic growth. The possibility of seeing wildlife actually aids economic growth. Denuding mountainsides and filling popular fishing rivers with sediment will eventually lead to a downturn in the economy.

"Increasing amounts of economic activity in [Montana] are landscape-related in the sense that economic activity is supported and enhanced by the high quality natural environment tied to these wild landscapes," writes Dr. Thomas Power, chairman of the Economics Department at the University of Montana in Missoula. "Sacrificing these economically important natural amenities in order to temporarily support an extractive industry in decline is the opposite of economic development. It is a prescription for ongoing economic decline."

While tourism is the most direct way of turning the region's scenic wealth into economic wealth, it is no panacea. In towns such as McCall, Idaho, and Dubois, Wyoming, recreation and tourism have flourished when old timber mills closed. But many of the communities that now depend on timber mills will find it hard to make the same transition to a recreation-based economy. Poor highway and airport access or the lack of decent streets, sewers or water systems render these communities incapable of attracting and managing a serious recreational economy. Loss of the timber industry will lead many small rural communities to the second stage of the old western cycle of boom and bust. Without additional capital development—both in public facilities and private businesses—even tourism won't replace the timber economies of towns such as St. Maries, Idaho, and Carson, Washington.

18

In areas where clearcutting has scarred the scenic beauty of the mountains, where streams run brown with mud from erosion, where bears, trout, birds and wolves are scarce, economic recovery may take decades. The proof lies east, back in the lands cut over by the lumber barons a century ago. Northern Wisconsin once looked a lot like northern Idaho, covered in towering white pines overlooking thousands of sparkling lakes and rivers. Once the trees were gone, wildlife species such as caribou, wolves and fishers disappeared, replaced by the prolific whitetail deer and other animals that adjust more easily to man. The giant pines and hardwoods were replaced by thick, brushy stands of aspen, called "popple" by Wisconsinites and only useful for making paper, and jack pine, a scruffy, thin pine with little economic value and hardly much more ecological value. Copper and iron mines closed early in the century, leaving northern Wisconsin among the poorest regions of the country.

The few islands of economic prosperity in northern Wisconsin are adjacent to those forests that were missed by the woodsmen, such as the Apostle Islands in Lake Superior, or the areas that have recovered the quickest, such as the lake country of Vilas and Sawyer counties. Ironically, among the most popular recreation and scenic areas in northern Wisconsin are small groves of old-growth trees that families such as the Weyerhaeusers protected as their own private wilderness parks. Brule River State Park in Wisconsin and the Sylvania Tract, a wilderness managed by the Forest Service in northern Michigan, remain much as they were when the lumber barons and steel kings of the Gilded Age of the 1890s locked them up. Today, national parks and federally designated wilderness areas preserve for future generations the Pacific Northwest's ecological jewels in a more ambitious and planned program of protection. These areas not only protect the future of the natural ecosystems and wildlife they contain, but also the economic future of the communities nearby. However, those towns surrounded by the cutover lands of Washington, Oregon, Montana and Idaho are doomed to share the fate of northern Wisconsin's rural hamlets as they await the next growth of timber.

But this scenario needn't be repeated everywhere. Millions of acres of roadless timberlands remain untouched adjacent to or near most of the rural communities of all four states. Not all of the timber has to be preserved to protect the natural ecosystems. In some cases the old lumber mills can be retooled so that they can produce new, higher-value wood products out of fewer and smaller-size trees. Though not true everywhere, owls, bears and loggers can live in harmony when the needs of all are considered. Before the Endangered Species Act was invoked to protect wild animals and their habitat, the needs of wildlife and even loggers were considered secondary to the desires of corporations, Congress and the federal land-management bureaucracies. The intervention of the Endangered Species Act in the cases of the northern spotted owl and Pacific salmon may enable the rural communities of the Pacific Northwest to avoid repeating the history of the upper Midwest.

The Endangered Species Act was not the first law Congress passed to protect endangered species. In 1964, the Bureau of Sports Fisheries and Wildlife, the predecessor of today's U.S. Fish and Wildlife Service, compiled the first list of endangered species, including sixty-three plants and animals. In 1966 and again in 1969, Congress authorized the Department of the Interior, which is ultimately in charge of the FWS, to make the list official and to begin protecting species on it by acquiring habitat. The original legislation had few penalties or enforcement provisions, and precious little habitat was protected.

In the heady days of environmentalism following Earth Day 1970, President Richard Nixon called for new, tough laws to protect the environment. During those first years of what Nixon called the "environmental decade," Congress passed the National Environmental Protection Act, the Clean Water Act and the Clean Air Act. It passed the Endangered Species Act in 1973 by a huge majority and Nixon signed it into law. The act's wide support was due not only to the wide popularity of such species as bald eagles and grizzly bears, but also to the fact that most senators and congressmen didn't know just how far they had gone.

One line, inserted as the act was drafted by the House Merchant Marine and Fisheries Committee, determined the character of the new law, setting it apart from every other environmental law ever written in the United States. In other laws, federal agencies are required to provide protection "where practicable." The Endangered Species Act requires agencies to take "such action necessary to ensure that the actions authorized, funded or carried out by them do not jeopardize the continued

existence of an endangered species." With that order, Congress elevated protection of all species to one of the U.S. government's highest priori-
20   ties. Unfortunately, its budget decisions have not reflected the same priority.

The strength of the law first became apparent in 1978. The U.S. Supreme Court ruled that if the building of the Tellico Dam on the Tennessee River jeopardized the survival of the tiny snail darter, then it could not be built. Specifically, the court ruled that Congress intended to place protection of endangered species above all other government activities.

Congress, recognizing the true power of the act, made it more sensitive to economic impacts in amendments added in 1978. Those amendments, which served to reduce the rigid nature of the original law, kept its absolute mandate intact. The one exception was the authorization of a cabinet-level committee, known now as the "God squad," that can decide to allow an activity to go forward despite its effects on endangered species. So far, the "God squad" has played only a minor role in endangered species decisions.

Only rarely has the Endangered Species Act stopped development or caused significant economic disruption. The snail darter and northern spotted owl cases are exceptions rather than the rule. And in the case of the snail darter, economic factors, not just the Endangered Species Act, killed the Tellico Dam. But the act has forced developers to consider the effects of their actions on endangered species and to modify proj-ects, in many cases increasing cost and complexity.

In 1990, the FWS examined 28,000 different activities, from timber sales to water projects, to determine if they would harm endangered species. Only two—a water project in Colorado and a harbor project in California—were stopped, and even they may eventually be allowed to move forward if the builders can make alterations that minimize the projects' effects on the species they jeopardize.

Two cabinet-level departments have authority to list species under the Endangered Species Act: the Department of the Interior, with its Fish and Wildlife Service, and the Department of Commerce, through the National Marine Fisheries Service. As of April 1992 there were 651 domestic species listed under the Endangered Species Act as either endangered or threatened. The FWS was responsible for 638 of the listed species and the NMFS was responsible for 19. If listed as endan-gered, a species is determined to be in immediate danger of extinction.

A listing as threatened means the species is likely to become endangered. The distinction as it affects management is small.

The act also authorizes the two agencies to protect critical habitat 21 and forces other federal agencies to consult with them before taking any action that could jeopardize a listed species. The act also contains penalties for the "taking" of listed species, which is defined broadly to include harassing, harming, capturing or killing. The decision to list a species must be made "solely on the basis of the best scientific and commercial data available."

Despite its ambitious goals, the Endangered Species Act has failed to prevent the process of extinction from occurring in the United States. According to Fish and Wildlife scientists, 7 species have been declared extinct after they were placed on the endangered list, and in the last ten years alone, 34 more species have become extinct while awaiting listing. Others estimate that up to 300 have actually gone extinct while candidates for listing. Another 3600 domestic species are listed as candidates for the endangered list, and the backlog continues to grow. Just to recover the species already listed would cost more than $4.5 billion, according to a Department of the Interior inspector general's report in 1990. Even with all of its power, the act falls far short of its lofty goals.

Saving the northern spotted owl is one of the act's most daunting tasks, since the owl's listing already has created widespread economic disruption in the rural, timber-dependent communities of Washington, Oregon and northern California. Plans drafted by the Forest Service and the Bureau of Land Management call for gradually reducing the average annual harvest of trees on federal lands in Oregon and Washington by 40 percent by the end of this decade. The recovery plan for the spotted owl could increase that reduction by another 40 percent.

John Turner, a Wyoming rancher who was appointed by President Bush as FWS director and who signed the decision to list the owl as threatened, had the unpleasant task of walking the point for the Bush administration's endangered species program. He had hoped to minimize the effects of owl protection on loggers and still protect the owl. Turner, Interior Secretary Manuel Lujan and others in the Bush administration challenged the Endangered Species Act's rigid requirements in attempts to allow larger timber harvests. Turner, who remained in the FWS post after Bill Clinton was elected president, had pushed for stricter measures than other Bush officials, but even his more protective plans were eventually overruled. The Department of the Interior and the Forest Service lost a series of court cases to environmentalists that, like the Tellico Dam decision of 1978, upheld the mandate of the act to protect the endangered owl at the expense of the timber economy.

22

Unlike the debate about the spotted owl, which has pitted logging jobs against owl protection, most of the players on both sides of the salmon debate agree there is room for both salmon and economic development. Yet finding that balance within the context of the Endangered Species Act will be challenging to the Pacific Northwest and the nation in the 1990s. In the last 100 years the region has chosen to press economic development with little regard for salmon. Now it is faced with the consequences of a century of overharvesting, pollution, dam building and destruction of salmon habitat.

The scope of the problem became clear during the 1980s. Easier answers to the diminishing numbers of salmon, spotted owls, grizzlies and other animals were still available. However, the Bush administration's intransigence in following the mandates of the Endangered Species Act continued a pattern started in the first days of Ronald Reagan's presidency. Reagan administration officials, Congress and land-management bureaucrats allowed short-term politics and economic considerations to override efforts to save species. The delay tactics and legal maneuvers resulted in precious habitat being destroyed. With ever-diminishing habitat left to work with, the task of restoring endangered species will be even harder. More limits on development and human activities will be necessary today and in the future because of our inaction a decade ago.

The history of the Endangered Species Act does include some success stories. Bald eagle and peregrine falcon numbers have risen sharply since they were listed as endangered in 1973. Most scientists agree the major factor in their recovery was the banning of the pesticide DDT, which was weakening the eggs of the birds and making reproduction difficult. Alligators, listed as an endangered species in 1973, are recovering in the South as a result of efforts to protect their habitat and to prevent poaching. Grizzly bear numbers have increased, at least in Yellowstone, since the great bear was listed as a threatened species in 1975.

The 1980s opened with many western states in open rebellion over control of federal land. In some states, including Idaho, Wyoming and Montana, more than half of the land in the state was owned and controlled by the federal government. Agencies such as the U.S. Forest Service, which managed the national forests, and the Bureau of Land Management, which controlled much of the desert rangeland, were 23 beginning to force their traditional tenants—ranchers, miners and loggers—to meet the bevy of new environmental laws approved by Congress in the 1970s. Several states remained bitter that the federal government had not handed over more land to them at statehood in the last century. And many land users, especially ranchers, viewed the federal estate as "their" land. The so-called Sagebrush Rebellion was quelled when Reagan placed in his administration several of its philosophical leaders, including James Watt, who was named secretary of the Interior in 1981. Indeed, many of the roadblocks placed in the way of those trying to protect endangered species in the 1980s were thrown up by former Sagebrush rebels or their staffs who went to work for the federal government. In many cases the federal courts stepped in and forced them to follow environmental laws.

In the 1990s, a new Sagebrush Rebellion is gaining a foothold in much of the region's rural areas, a backlash to court decisions and the growing influence of the environmental movement in the West. The so-called wise-use movement is made up of loggers, miners, motorbikers, farmers, ranchers and snowmobilers who see their lives threatened by the land-protection measures included in the Endangered Species Act. These groups are banding together and becoming a new political force. The foot soldiers of this movement are frustrated with the control government has over their lives. In the Pacific Northwest, most of the land surrounding most rural communities is federally owned. The residents of these communities cut timber, graze their cattle, mine and play on public land. For decades, their land-management practices were unquestioned. But in the last fifteen years a host of environmental laws and a growing interest from urban areas has brought more players into the management picture. The residents of cities such as Boise and Portland want to protect the hinterland both for their enjoyment and for its intrinsic value.

Rural residents of the Pacific Northwest and northern Rockies want to hold on to the traditional lifestyle they have built around their professions and recreation. Environmental groups have appeared to many loggers, mill workers, miners, ranchers and small-town business people to care more about animals than about the future of rural populations. Traditional industry groups, such as the Intermountain

Forest Industry Association, have been looking out for the mill and mine owners, so rural residents formed groups such as the Oregon Lands Coalition, Citizens for Multiple Use in Idaho and Montana, People for the West, and the Blue Ribbon Coalition, which represents motorized recreationalists. Large mining companies, the timber industry and motorcycle manufacturers, having lost some of their political clout in Congress and needing to expand their constituencies, have heavily funded some of these groups, convincing them that their agendas and industry's are the same. The result has been further polarization between urban and rural public-land users, who share many of the same values but clash over the future of extractive industries such as mining and logging.

Luther Probst of the Sonoran Institute, a community-planning company in Tucson, Arizona, that works toward a future based on consensus among all groups in a program called Successful Communities, compares the wise-use movement to the ghost dancers in Indian tribes a century ago. The ghost dancers believed that if they danced the ghost dance they would become invincible and could easily drive the white men away. The modern ghost dancers—members of the wise-use movement—fill meeting halls and attempt to challenge federal laws with local ordinances, hoping to reverse the changes taking place throughout the West. The modern ghost dancers, like the Indians of the nineteenth century, will not stem the tide of a growing urban population more interested in protecting the region's precious wild creatures and land than in preserving the traditional lifestyle of its rural residents.

John Turner says what is needed is a new strategy to protect species before their populations fall to the point that they are eligible for listing. "If we do a better job of monitoring species that are starting to get into trouble, we do a better job of preventing species from getting to a crisis state," he says. It would cost less to prevent a crisis than to clean up afterward, and we may have a better shot at preserving more species.

In fact, what is really needed is a holistic approach that looks at entire ecosystems and all of their various parts. In his agency's strategic plan, Turner had proposed such an approach and had set a new goal: protecting biodiversity. The word, short for biological diversity, means the sum total of life, ecosystems, species and gene pools. Many conservation biologists argue that it is necessary to preserve diversity in each of those fundamental sets in order to prevent a decline in the biosphere that could ultimately undermine the quality or even the survival of human life. Any loss of diversity in any part of the natural world is considered by many scientists to be harmful to the entire planet.

E. O. Wilson, a renowned entomologist at Harvard, says that humans came into existence at a time of the greatest biological diversity in the history of the world. The expansion of human population and the alterations it has made to the natural environment are reducing biodiversity to its lowest level in 65 million years, Wilson has written in *Scientific American*. "The ultimate consequences of this biological collision are beyond calculation and certain to be harmful," he notes.

According to Wilson, 1.4 million species have been identified and given scientific names, with conservative guesses placing the planet's total number of species at 4 million or greater. Terry Erwin of the Smithsonian's National Museum of Natural History estimates that when the insects and other arthropod species of the tropical rain forests are factored in, as many as 30 million species may exist. Wilson says conceivably there may be as many as 100 million species. According to the World Wildlife Fund, species are disappearing worldwide at a rate of about 150,000 per year—that is, seventeen every hour. Of course, many of these species would be lost naturally, without the influence of human activity. The loss of others may be biologically insignificant. But as Wilson points out, the effects of the loss of a species are both unpredictable and irreversible. The consequences of the loss of one species or a hundred species cannot be wholly estimated because much of the world's flora and fauna have not been studied, and those that have been may not have been studied for the values that may ultimately be discovered. And those values may well be tied directly to human needs and desires or to the ecosystem in which the species lives.

"The last word in ignorance is the man who says of an animal or plant: 'what good is it?' If the land mechanism as a whole is good, then every part is good, whether we understand it or not," wrote Aldo Leopold in his classic essay "Round River" in *A Sand County Almanac*. "If the biota, in the course of aeons, has built something we like but do not understand, then who but a fool would discard seemingly useless parts? To keep every cog and wheel is the first precaution of intelligent tinkering."

Take, for instance, the Pacific yew tree, a species of the old-growth forests of Idaho, Oregon and Washington. This needled evergreen tree grows in the shadows of old-growth Douglas firs throughout the region. Until a few years ago, it was viewed by the Forest Service as a weed, a useless victim of the clearcutting frenzy of the 1980s.

Today, we know that the yew's reddish purple bark is the only source of a substance called taxol. This naturally derived drug has reduced tumors in some patients with ovarian cancer, a malignancy that is resistant to other forms of chemotherapy and that kills 12,400 women a year. Environmentalists and cancer researchers tried to get federal officials to list the yew as a threatened species in 1991, but failed. Even now that a synthetic taxol has been developed, the yew tree was the key to the cancer breakthrough. Its medicinal use would not have been discovered had it become extinct.

The yew tree is not an exception. About half of the nation's prescriptive medicines come from living species. These naturally derived medicines are successfully fighting such diseases as leukemia, herpes, encephalitis and high blood pressure. Farmers also benefit from biodiversity, both in seed crops and, ironically, in combating insects. Agricultural scientists use germ plasm from wild species of corn, wheat and soybeans to maintain and expand crop productivity, to fight new insect pests and to aid crops in adjusting to climate changes and other threats to the environment. Insects that feed on other insects also are replacing pesticides for pest control. Even though humans depend on fewer than twenty plants for 90 percent of our food consumption, about one-third of the earth's 250,000 plant species are believed to be edible, according to a National Academy of Sciences report. Several hundred of these, the report states, could be used to relieve world hunger and to improve nutrition. With the threat of global warming and consequent major climate changes, we may be losing valuable species at a time when we will need them to adapt to future conditions. "Crushed by the march of civilization, one species can take many others with it, and the ecological repercussions and rearrangements that follow may well endanger people," says the Worldwatch Institute, a nonprofit research organization, as noted in a 1992 report of the U.S. General Accounting Office.

The Endangered Species Act is the best protection the United States has for preserving biodiversity. It still falls short of protecting every genetic variation that occurs in nature, and the debate over how far it should go to preserve the gene pool continues. For years scientists have been debating how to delineate species. The act has only heated up that debate and brought into it special-interest groups on both sides. The act provides protection for endangered species, subspecies and "any species which interbreeds when mature." How far the act will go to protect subspecies or specific populations of species may be decided in the Pacific Northwest, where salmon will be a major focus.

In the Northwest, at least 420 distinct stocks of salmon and steelhead once migrated up and down thousands of miles of tributaries throughout the Columbia River Basin. Each stock had its own genetic blueprint—characteristics developed over thousands of years of evolution that made it uniquely suited to the tributary where it spawned, the rivers through which it migrated and the section of ocean in which it lived. Scientists estimate that there are about 200 separate stocks of salmon and steelhead remaining in the Columbia River and its tributaries. Another 220 stocks disappeared in the last 100 years, according to Oregon Trout, a salmon advocacy group. The American Fisheries Society has identified 195 naturally spawning stocks of salmon and steelhead in Oregon, Washington, Idaho and California that are in some danger of extinction. Ninety are identified as high risk.

Pacific Northwest residents face the real prospect of the loss of Pacific salmon in this decade in the Snake River. And by the end of the next century the miraculous runs may have disappeared throughout the region. Grizzly bears and caribou also may disappear. The decisions made before the end of this century may well decide whether many species, large and small, live to see 2100.

In the twenty-first century, the challenges will grow both in the Pacific Northwest and in the world as a whole. The world population is expected to double to 10 billion by 2050. A five- to tenfold increase in land development activity may be needed to meet the basic needs of this future human population. Earth's natural systems already are stressed, and even conservative scientists such as E. O. Wilson raise serious questions about how these systems will respond to such population growth alone. In the United States, the population is estimated to grow by as much as 40 percent to 340 million over the next fifty years. The pressure on all of its natural resources will increase.

Grizzly bears, wolves and salmon have remained in the Pacific Northwest and the northern Rockies primarily because of the region's isolation. But the Pacific Northwest no longer is as secluded as it once was, and the competition for space and resources will become ever more intense. Much of what has been lost in the ecosystem during the

28

last century has been the result of ignorance. Now, the piece-by-piece dismantling of the natural ecosystem of the Pacific Northwest would happen with our full knowledge. We in this generation won't have the right to say we didn't know. Aldo Leopold, the ecologist who almost single-handedly created the science of wildlife management, wrote in "Round River" of the penalties of an ecological education: "One lives alone in a world of wounds."

Leopold also wrote that "much of the damage inflicted on the land is quite invisible to laymen. An ecologist must either harden his shell and make believe that the consequences of science are none of his business, or he must be the doctor who sees the marks of death in a community that believes itself well and does not want to be told otherwise."

When those words were published in 1953, Leopold was nearly alone in his thinking; only a few understood as well as he the folly of random tinkering with the ecosystem. But today each of us must be the doctor or we all will soon be the patients.

Thomas Berry, a theologian and environmental philosopher, writes in his book *The Dream of the Earth* that more than simple changes in strategy are necessary to preserve the natural communities in each region and the world as a whole. We must redefine progress.

"The difficult transition to make is from an anthropocentric to a biocentric norm of progress," Berry writes. "If there is to be any true progress, then the entire life community must progress. Any progress of the human community at the expense of the larger life community must ultimately lead to a diminishment of human life itself."

2. Living with Grizzlies

THE grizzly bear sow accelerated to a gallop as she left the timber and caught the elk calf running. She dragged and carried the kicking calf toward Antelope Creek in the northeast quadrant of Yellowstone National Park. Elk cows milled around, sniffing to see if it was their calf the bear had grabbed. The sow, who wore a radio collar and was tagged with the number 59, dropped her prey and rounded up her cubs, which she had left behind in the chase. Later she returned to carry the calf into the protection of the woods for a much-needed spring feast for her and her cubs.

The sow's success at elk predation, captured on film by bear researchers Steve and Marilyn French, independent bear researchers with the Yellowstone Grizzly Foundation, demonstrates a dramatic change in grizzly behavior that took place in the 1980s in the Greater Yellowstone Ecosystem, the large, still mostly wild area that includes the world's first national park and its surrounding mountains and valleys. Instead of grubbing through garbage dumps as earlier generations of Yellowstone bears did, Bear 59 and others learned to hunt the park's growing elk herd and to fish for the increased cutthroat trout population that run up the streams around Yellowstone Lake. In addition to their traditional diet of roots, forbs, berries and seeds, Yellowstone's opportunistic grizzlies are even meeting seasonal nutrition needs by eating ants and moths.

Unfortunately, Bear 59's life story also bolsters the fears of environmentalists who worry that human-caused mortality continues to threaten the future of the grizzly. While Bear 59 had learned to prey on elk, she also had become habituated to humans. She was often seen

29

near the roads around the park's Canyon Village development and drew a crowd of amateur photographers anytime she came into view. In 1986, one photographer, William Tesinsky of Great Falls, Montana, got too close. She killed and ate him. Soon after, at the hands of park rangers, she too was killed. Since her cubs already had mysteriously disappeared, her eight years of learning to cope with Yellowstone's environment was not to be passed to another generation.

30

Fortunately, fewer bears are succumbing to death resulting from confrontations with humans than they did a decade ago. The combination of reduced mortality and dramatically increased sightings of grizzlies, particularly adult females with cubs, has made the agencies that manage grizzly bears and many researchers optimistic about the immediate future of the huge brown bears that have become a symbol of wilderness in the West. "We think things will probably improve for the next few years unless something unforeseen or catastrophic happens," says Richard Knight, the Interagency Grizzly Bear Study Team leader.

Knight, the gruff, direct-speaking, grizzly research team leader for the National Park Service, estimates that at least 228 bears now roam the Greater Yellowstone Ecosystem in western Wyoming, southern Montana and eastern Idaho. Chris Servheen, grizzly bear recovery coordinator for the U.S. Fish and Wildlife Service (FWS), said the number of grizzly sows with cubs rose from 30 in 1983 to as many as 60 by 1992.

These figures had brought a stream of federal resource officials to Yellowstone to crow over their success at restoring the grizzly bear. Many in the Bush administration, from former Interior secretary Manuel Lujan down to Knight, wanted Endangered Species Act protection removed from the grizzly in Yellowstone and in the Northern Continental Divide Ecosystem, which includes Glacier National Park, the Bob Marshall Wilderness and other lands surrounding the natural areas of northwest Montana. To date, the Clinton administration has made no major changes in the direction of grizzly management, and it is unclear whether they will continue to push for delisting or take a more cautious approach. Bears were listed as a threatened species in 1975, and ever since, activities such as logging, mining, oil drilling and recreation have been restricted on large areas of public land. Declaring success and officially delisting the grizzly bear would ease the restrictions that have caused many in the region to resent the bear and the government controls it carries with it.

Yet despite the clearly upward trend of the Yellowstone grizzly population, many environmentalists and researchers remain pessimistic about the future of the grizzly in Yellowstone and throughout the northern Rockies. Even if the odds are good that the bears will survive there for the next fifty to a hundred years, the picture gets murky as scientists gaze beyond that. At the same time that intensive management has reduced mortality and allowed more bears to reproduce, intensive human development has been eating away at the once desolate land needed as a home by the bears. Loggers have clearcut thousands of square miles of once pristine wilderness. Roads have been pioneered along rivers and streams, carrying hunters and other people deep into the heart of bear country in densities that grizzlies simply can't tolerate. Miners are digging up entire mountains where grizzlies used to find important seasonal foods. Oil and gas exploration is pushing bears away from key migration routes to other high-quality habitat. Subdivisions of second homes and vacation retreats are filling the remaining valleys that link the fragments of wild land left for the bear. In short, humans are crowding grizzly bears out of the last tracts of livable space left in the continental United States.

That competition between humans and bears for the same space is the major hurdle to long-term preservation of the grizzly bear in the Lower 48 states. It is the essence of the conflicts over the Endangered Species Act nationwide. If we cannot preserve the grizzly bear in the northern Rockies, then it is doubtful humans can stop the plummeting level of biodiversity on earth. Most of the Pacific Northwest was introduced to the Endangered Species Act with the controversy over the northern spotted owl that began in the late 1980s; yet in the eastern portion of the region—the northern Rockies—ranchers, loggers and others who share space with the last grizzlies in the Lower 48 states have learned to live with the law for more than fifteen years. The relationship between that area's residents and those who have used the law to preserve grizzlies has been as rocky as the peaks that dominate the skyline. And the future of the bear and human development in the region remains uncertain. Grizzlies have a limited capacity to adjust to humans, and the human capacity to live with an often ornery neighbor is challenged regularly in this region.

The example of how grizzlies and humans have lived together since 1975 shows both the strengths and weaknesses of the Endangered Species Act and the institutions in place to carry out its mandates.

Grizzly bear numbers continue to rise, yet their habitat continues to shrink. The experience of the last fifteen years shows how easy it is to protect the species themselves but how hard it is to protect their crucial habitat.

32

When Lewis and Clark entered the northern Rockies in 1805, there were about 100,000 grizzlies spread throughout the West, Fish and Wildlife Service biologists estimate. Today federal managers are trying to manage 14 million acres of federal lands to preserve about 1000 grizzly bears—the remaining grizzly population. The controversy concerning bear recovery didn't reach the crescendo of the owl debate partly because most of the time when conflicts between humans and bears arose, it was the bears that were moved and not the humans. Those who predict dire economic problems from protection of endangered species ignore the case of the grizzly bear. Even in the face of continuing evidence of a long-term threat to the bear's future, efforts to protect the bear have rarely curtailed human development. The question is whether managers have been too willing to allow development at the expense of precious bear habitat.

Whether the bear has been pushed too far, however, will likely be decided in the Greater Yellowstone Ecosystem, the southern edge of the grizzly's range. The 9500 square miles of land in and around Yellowstone National Park protected as grizzly bear habitat is small, and the isolated population of Yellowstone bears may remain in a precarious state indefinitely.

"Optimism about the long-term viability of the Yellowstone grizzly bear population is not warranted," said researchers David Mattson and Matt Reid in a landmark paper published in 1991.

Mattson, the young, respected assistant to Richard Knight, made a public break with his boss when he published the paper with Reid, a long-time grizzly bear advocate. Mattson acknowledges the short-term improvements of Yellowstone's bears, but is skeptical that there is enough habitat protected to ensure the viability of the bear population. "Nobody has proven there's enough habitat out there to support viability in a meaningful sense. I think the burden of proof should be on us to prove that," Mattson says.

Knight, who has headed the Yellowstone grizzly research team since 1973, says there is enough habitat to protect Yellowstone grizzlies— but just enough. In public forums, such as the International Conference on Bear Research and Management in Missoula, Montana, in

1992, Knight has avoided specifics. The battle-worn biologist has had his words used against him many times in the past from both supporters and opponents of grizzly bears. He particularly hates to 33 guesstimate the current population of bears in Greater Yellowstone, preferring instead to offer a minimum population estimate that he acknowledges is a ball-park figure.

Other bear biologists are increasingly critical of Knight's population estimation method, his lack of concern for long-term bear survival issues and his offhand manner of responding to queries. However, the federal agencies that manage the land in the Greater Yellowstone Ecosystem have displayed confidence in his judgment. When Knight told them in the early 1980s that the Yellowstone bear population was in a serious state of decline, they initiated strict limits on grazing, logging and recreational use to turn the trend around. Those actions, although not popular at the time, were successful, and now Knight offers the optimistic view that Yellowstone has enough bears to remove them from the "threatened" list. This has especially endeared him to those land managers desperately seeking a success.

Controversy over grizzly management in Yellowstone is not new. John and Frank Craighead studied Yellowstone's bears from 1959 until 1972, when their opposition to the abrupt closing of Yellowstone's garbage dumps got them kicked out of the park. Their dispute with former Yellowstone research chief Glen Cole polarized the biological community for nearly a decade. Cole closed the dumps because he and others in the National Park Service believed that the bears' use of the dumps was unnatural and therefore out of step with the new natural management program the agency had begun. The Craigheads, while not totally opposed to closing the dumps, believed that the entire bear population was at least partially dependent on the dumps for food and that it had to be weaned off the century-old food source slowly. That fight, although primarily revolving around the dump issue, also drew lines over the issue of population estimates and habitat needs of bears. In the 1960s, the Craigheads had marked 264 bears in Yellowstone and said this represented 75 percent of the bear population in the entire ecosystem. They estimated a population of 245 bears in 1967. Cole believed there were 50 to 100 "backcountry" bears that had not been identified at the dumps by the Craigheads and that they had marked less than the 75 percent they claimed. Cole believed there were 300 to 350 bears in the ecosystem. Even though the Craigheads' population numbers are the more widely accepted estimates today, the debate about how many bears Yellowstone can support still rages.

34   Yellowstone has captured most of the attention concerning grizzlies because as the first national park in the world, it is watched closely as an example and a measuring stick for conservation efforts worldwide. Yet the fate of the grizzly rests not simply on Yellowstone but on the entire northern Rockies. Almost none of the biologists who study Yellowstone believe that the recovery area is large enough to sustain bears without interaction with other bear populations in the northern Rockies. Some scientists believe as many as 35 million to 42 million acres of land are necessary to the survival of grizzlies.

Today, only about that much land in the northern Rockies of the United States and Canada remains usable grizzly bear habitat. So to meet that upper estimate, all of the remaining habitat would have to be protected and the scattered pieces of habitat eventually connected through stricter management controls. For while several large, desolate pieces of land remain scattered around the northern Rockies, they are miles away from one another. The Craigheads and other scientists argue that these biological corridors, or "linkage zones," are key to the long-term survival of the great bear in the Lower 48 states. That is a significantly taller order than preserving the bear in Yellowstone. In the smaller pockets of wild country where bears still roam, such as the Cabinet-Yaak Mountains of northern Montana, Idaho and southern Canada; the Selkirk Mountains of northern Idaho, Washington and British Columbia; the Selway-Bitterroot Wilderness in Idaho; and the North Cascades in Washington, grizzlies hold on only by a thread. Only in the Northern Continental Divide Ecosystem, which surrounds Glacier National Park and runs into Canada, are grizzly numbers higher than in Yellowstone.

The Yellowstone bears face all the problems their relatives face throughout their limited range. Oil and gas development, subdivisions, recreational developments, road building, logging and mining compete with the grizzly outside Yellowstone National Park. The more than 3 million visitors and developments in prime grizzly habitat, such as Fishing Bridge, Grant Village and Lake Hotel in the park, make much of the great bear's prime habitat off-limits for grizzlies despite the dictates of the Endangered Species Act. Federal and state managers keep trying to balance the different uses and mitigate their effects on the bear. But only rarely have they actually halted the decline of bear habitat. Their biologists continue to warn them that only so much use can be allowed without squeezing out the bears. "We can't shoehorn every use into the Yellowstone ecosystem," Chris Servheen says. "You can't high-tech every use and protect the bear." Unfortunately, federal officials continue to try to do exactly that.

35

What makes grizzly bears such a challenge to manage is that they themselves defy management. Grizzly bears are fierce, intelligent, wild animals that refuse to compromise. Their violent nature is a defining character trait necessary for protecting their young from other predators, since they spend little time under the cover of forest.

Grizzly bears are larger than black bears and can be distinguished from their cousins by their longer, curved paws, humped shoulders and concave facial features. Occasionally a male will weigh up to 1000 pounds, but grizzlies in the northern Rockies average from 400 to 600 pounds for males and 250 to 350 pounds for females. Adults stand about four feet at the hump when on all fours and up to eight feet upright.

Grizzlies can live up to nearly fifty years but in Yellowstone are more likely to live about twenty-five years. A female in the Cabinet Mountains was thirty-seven in 1992. Females don't begin breeding until they

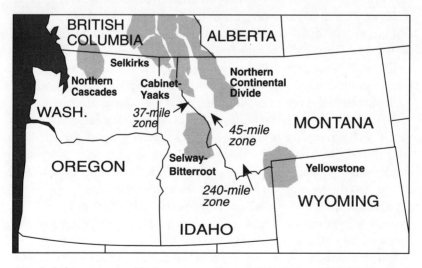

Grizzly habitat areas and linkage zones. (Source: U.S. Fish and Wildlife Service.)

are at least three years old, and then breeding takes place an average of every three years. Litter sizes vary from one to three cubs. Grizzlies have one of the lowest reproduction rates of any land animal, a factor that plays an important role in the recovery of the species. In short, recovery is generally slow and subsequently precarious and heavily dependent on reproducing females.

Sows with cubs are also the most dangerous bears, since any encounter likely triggers a violent defensive attack by the sow. This not only places the human in danger, but the bear too, since most attacks, like that of Bear 59 on William Tesinsky, end up in the bear's death as well.

Bears need space—wild, undeveloped and open. Biologist Mark Shaffer estimates the mean average area requirement of a single grizzly bear in the northern Rockies to be thirty-three square miles. That would mean the Yellowstone recovery area can support roughly 288 bears. Unfortunately, human development such as roads, resorts, mines and clearcuts displace bears and make much of the habitat in Yellowstone unusable by its grizzlies.

Bears are omnivorous, opportunistic feeders that specialize seasonally on the most abundant food sources. Most of the year roots, berries and leafy plants make up the bulk of the grizzly's diet. But other foods, such as elk meat and trout, are richer in protein and therefore very important, especially in the post-denning period, when bears need the most food. Bears also need additional protein and carbohydrates in the fall to build up weight and strength before they hibernate in November. During the period from early August through November bears go through a process known as hyperfacia, during which they have a particularly large appetite, and if food sources such as whitebark pine nuts are not available, they must seek out alternatives.

A good year for bears in terms of food is usually the same as a good year for the irrigating farmers who grow potatoes on the southwest edge of the grizzlies' range. A wet, cold winter leaves heavy snows in the mountains, which improves vegetation, kills elk and buffalo for spring grizzly feeding, and generally enhances the bears' food base throughout the year. Whitebark pines produce high-fat nuts that are extremely important to bears in the fall. Squirrels collect caches of the nuts and the opportunistic grizzlies steal them.

Problems arise in poor food years, when grizzlies need more range to survive. A poor year is one that is dry, like the five out of six drought years from 1985 to 1990, which created the conditions that spawned the great Yellowstone fires of 1988. Bears find fewer winter-killed elk and buffalo, and vegetation is sparse and dry in the late season. When a poor year coincides with a whitebark pine nut failure, which happens about once every four years, bears will wander widely in search of other food. Often this draws them toward humans and trouble, usually in late summer, when their usual food sources dry up.

37

Before 1971, when the garbage dumps were closed in Yellowstone, the bears depended on the park's garbage to get them through the lean seasons. But when the dumps closed, the bears reverted to the next best thing: campground food and garbage dumps in surrounding communities. Management efforts have largely cleaned up these sources, and bears have been forced to adapt. Research by the Interagency Grizzly Bear Study Team indicates that the bears began using new natural food sources and finding ways to use traditional sources longer by the early 1980s.

Grizzlies are different from many animals in that they learn most of their survival skills from their mothers, rather than acting on instinct. This fact makes protecting mother bears even more important. The park's grizzly population boom in the late 1980s came at the same time bears were relearning natural feeding patterns. One could argue that the two are connected, since drought dried up many food sources and good nutrition is necessary to healthy reproduction. Most important, perhaps, has been the bears' renewed ability to feed on elk in the spring and in the fall. Yellowstone's elk herd has grown from about 4000 in the 1960s to more than 30,000 by 1990, providing bears with a substitute for garbage. "We have seen a trend toward increasing use of elk in the summer and fall," says David Mattson. "It's not just increased scavenging but also an increase in predatory behavior."

Many bears have learned to kill elk calves regularly in the late spring and summer, and some have learned a more complicated fall hunting strategy of preying on rutting bulls. The Yellowstone elk population, which faces little hunting pressure, has a large number of big, older bulls. Many compete for the same cows during the breeding season, and herding their harems and fighting off competitors leave many bulls very weak after a few weeks. Grizzlies have learned to wait for the right moment to kill one of these bulls with little effort.

It is complicated behavior based on the spring experience of preying on extremely weak animals, according to Mattson. Bears had learned to kill elk that were on the verge of death in the spring and then took that experience to the next level to kill the weakened bulls. In poor food years, such as 1988 and 1992, bears even were seen killing relatively healthy elk.

Just as elk numbers have risen, so have cutthroat trout in Yellowstone Lake. Trophy fishing regulations have increased once-depleted cutthroat numbers to stream-choking proportions during the spring. Today the trout have become an important summer food source for bears. "We estimated that there are as many as forty different grizzlies using Yellowstone Lake spawning streams during June and July, so it's a significant eating pattern," says Mattson.

While eating elk and trout are the most obvious relearned feeding habits, the grizzlies also are increasingly eating ants and cutworm moths in alpine areas. Scientists are only beginning to learn how important these sources are at critical times of the year when other sources are less available. The moth feeding is attracting as many as ninety-five grizzlies annually to ten sites in the late summer, according to bear researcher Steve French. French says the newly observed behavior is important to the bears because they are getting a concentrated high-energy food during a season when other food sources are scarce. It's also important to researchers because it gives them a chance to study bear social behavior, which has been harder since bears dispersed into the backcountry in the early 1970s. "We haven't seen this cluster of bears since the garbage dumps closed," says French.

The bears feed on the moths from July through early September on the upper slopes of peaks as high as 12,000 feet in Wyoming east of the park. French says researchers saw fifty-one bears on four of the ten sites in one day.

The grizzly's major food sources remain succulent grasses, forbs, leafy plants and roots. These foods are available most of the year, and because of the grizzly's digestive chemistry, are turned into the fat the bears need as easily as meat, fish, moths and berries. But many of the traditional feeding areas for bears are no longer available to them. In the Targhee National Forest in Idaho, for example, where more than 165 square miles of forestland was clearcut in the 1980s, a seasonal feeding area that once was regularly used has become marginal.

Some researchers say our unwillingness to provide enough habitat makes supplemental feeding one alternative to expanding the bears' range, particularly in a poor year. Frank and John Craighead, who did much of the original research on Yellowstone's grizzly bears, say most healthy bear populations have "ecocenters" such as salmon spawning streams that provide feasts for bears. Yellowstone has not had ecocenters since the dumps closed, says Frank Craighead. He has called for the development of artificial, backcountry ecocenters, using the abundant elk as food. "One of the reasons for establishing ecocenters is that it will cause a lot less friction among the users," says Craighead. "It will be beneficial, when the population increases, in concentrating grizzlies away from people."

But Mattson says the problem would be targeting the ecocenters to female grizzlies with cubs, bears that would need the extra nutrition in poor years. Older, dominant males, says Mattson, get the best food sources in the ecosystem and force females and younger males to poorer feeding areas. Moreover, the increased elk and cutthroat numbers provide enough additional food for the bears to allow them to survive, at least in the good years, according to Richard Knight.

Grizzlies are currently using most of the habitat in the Yellowstone region that humans are willing to share. In addition, they have ranged as far south as Spencer Mountain just north of Swan Valley, Idaho, and are recolonizing the Wind River Range in Wyoming, more than 100 miles southeast of Yellowstone. The increased numbers of wide-ranging grizzlies are running into humans, usually at the expense of the bears. In 1990, the Greater Yellowstone Ecosystem lost eight bears, including four adult females. Hunters killed four, two others were electrocuted and another was hit by a car. Park rangers captured an eighth bear and sent it to a research lab, which is the same as a mortality in terms of the population.

The rising number of grizzly bear killings, especially at the hands of elk hunters around Yellowstone National Park, is a reminder that Yellowstone's grizzlies are trapped on an island surrounded by a sea of development. Bear and hunter conflicts and continued development jeopardize grizzly range and could keep the great bear listed as a threatened species for years to come.

Even after the grizzly was listed as a threatened species in 1975, hunters in the Yellowstone region have continued to shoot and kill grizzlies. Some have been poached, and others were killed because of mistaken identity or in self-defense. Within a one-month span in 1990, for example, Fish and Wildlife Service officials said four grizzlies were shot in the Wyoming wilderness southeast of Yellowstone.

In one of the incidents, Ed Higbee, a fifty-two-year-old real estate broker from Cody, Wyoming, was pushing through heavy timber in the remote Thoroughfare country in the Teton Wilderness southeast of Yellowstone on September 12, 1990, with six other hunters. Late in the day Higbee was working his way uphill on a narrow trail. Suddenly, a five-year-old grizzly sow jumped from behind a log and charged Higbee. He got one shot off from the hip, striking the growling bear before she pounced on him. Jeff Meyer, one of Higbee's hunting partners, heard the commotion and ran uphill to find the bear mauling the former Cody Chamber of Commerce president. Meyer shot the bear once, and it stalked off and died, leaving a cub in the woods.

The Higbee incident got a lot of publicity in regional newspapers, and some officials say it started a chain reaction. As a consequence, many hunters entering prime grizzly habitat to hunt elk were frightened, edgy and prepared to shoot a bear, according to John Talbott of the Wyoming Game, Fish and Park Department. "We have a lot of people entering the woods prepared to shoot a grizzly on sight," he said.

While hikers have been taught to make noise in grizzly country, hunters move around quietly to avoid scaring off their quarry. That makes hunters more apt to surprise a grizzly, says Jim Klett, an FWS special agent from Jackson, Wyoming. The killing of Higbee's attacker was ruled justifiable, but the three other bears that died that year might not have had to, officials said. One was found hit in the back of the shoulder. "It was a close encounter, not a direct attack," Klett says. "I don't blame the bear." But he says hunters might feel threatened in such a situation and are given the benefit of the doubt. It remains one of the gray areas in enforcing the Endangered Species Act for grizzlies. "Proximity to a 600-pound bear at four to eight feet triggers a response in people," he says.

Richard Knight, who has spent many hours in close encounters with bears as a researcher, is less willing to give hunters the benefit of the doubt. He believes hunters that enter grizzly country should be prepared to run into grizzly bears occasionally without feeling the need to kill them. Those hunters who kill grizzlies aren't macho characters to Knight, who simply calls them "wood wimps." "If they're afraid of bears they shouldn't hunt in grizzly country," he says.

Despite the problems with hunters, managers are considering a special "nuisance grizzly" hunt in the Yellowstone ecosystem aimed specifically at problem bears, thus allowing hunters to bag grizzlies that would have to be trapped or removed from the wild anyway. "If it did happen, it would be very rarely implemented," Chris Servheen says, adding that a similar nuisance grizzly hunt was on the books in Montana, where in four years just two grizzlies have been killed.

41

The Fish and Wildlife Service's grizzly recovery zone includes portions of the five national forests that surround Yellowstone. Inside the boundary of the recovery zone, grizzlies are afforded special protection. Outside the zone, grizzlies still cannot be legally killed, but their human neighbors don't have to make room for them, either. Human-bear confrontations are resolved by moving the bears.

Despite the special protection of the recovery zones, bear deaths at the hands of humans are a problem throughout grizzly country. Grain train derailments along the Middle Fork of the Flathead River in northern Montana attract grizzlies from miles around, resulting in some of the bears being struck by trains. In the Cabinet Mountains of northwest Montana, hunters continue to shoot bears by mistake. In one case, a hunter said he mistook a grizzly cub for an elk. Along the Idaho–British Columbia border, a labyrinth of logging roads brings hunters and huckleberry pickers into what otherwise would have been remote grizzly country.

Educational programs and increased enforcement in the late 1980s and early 1990s cut bear deaths in the Selkirks dramatically, says Wayne Melquist, the Idaho Department of Fish and Game nongame and endangered species coordinator. Researchers have trapped and tagged about forty bears in the Selkirks since 1983, but bear numbers may have reached the carrying capacity of this mountainous border area in Idaho, Washington and Canada. "We may be at the point that we can't maintain more bears," says Melquist.

An unprecedented timber harvest throughout Montana has fragmented much of the best bear habitat in the Northern Continental Divide Ecosystem. Conservationists such as Lance Olsen of the Great Bear Foundation worry that the bears in and around Glacier National Park are worse off today than they were when listed in 1975. "We've lost an awful lot of prime habitat since 1975," says Olsen.

In 1990, eleven grizzlies were removed from the Northern Continental Divide, including four wild bears that were sent to research labs or zoos. Two bears were struck by trains and killed in 1989, and four others were killed by trains in 1990 at a derailment site. A fifth bear was struck by a train in 1990 but the death was unrelated to the grain train derailments. Burlington Northern trains derailed along the Middle Fork of the Flathead River in 1985, 1988 and 1989, leaving more than a million pounds of corn along the tracks. A federal report documents that fourteen to sixteen different grizzly bears became dependent on the corn at the derailment site and were unafraid of the trains or human activity.

Meanwhile, residential lot sales along the North Fork of the Flathead River threaten to remove even more habitat from the heart of grizzly country. Logging roads snake through most of the national forests of northern Montana. Olsen says that had federal agencies followed the Endangered Species Act requirements, much of the habitat lost since 1975 would still remain. Critical habitat would have remained protected on private lands as well as on public land.

Instead of designating critical habitat, federal officials developed the Interagency Grizzly Bear Guidelines. The guidelines were written by the Interagency Grizzly Bear Committee, a group of state and federal land and wildlife managers that coordinates the bear's management in the Lower 48 states. The committee sets regulations to keep human-bear conflicts to a minimum on public lands in the West. It teaches federal land managers how to resolve conflicts and determines how development should proceed in grizzly habitat. In the Yellowstone ecosystem, the same land managers have set up situation areas to provide protection for the bear.

In a Situation 1 area, the bear is the dominant occupant of the area and all other users are to be considered second to its needs. In Situation 2, bears are considered infrequent visitors and get equal consideration with humans in management conflicts. Situation 3 areas are usually on private land, and bears can be moved immediately if problems arise. Tony Povilitis, a college professor from Colorado who has been a frequent critic of bear management, believes that the situation areas were set up more for political reasons than for biological ones. Povilitis contends that the boundaries were drawn specifically to leave commercial timber stands and grazing allotments outside of Situation 1, even though bears were occupying these areas and they were important bear habitat. "It's hypocrisy to say it's based on biology if it's based on politics," he says.

Servheen and others argue that all the regulations in the world won't protect the bear if westerners don't support bear management. "I think the biggest problem for the grizzly bear is the attitude of the local people," says Bob Hammond, Shoshone National Forest Crandall District ranger. Integrating bear management into the lives and occupations of the human residents of bear country is a great challenge, but it is necessary if bears are to survive long into the future. The human population of the region is not going to drop, and with more people will come more pressures on the bear.

Some residents, such as lumberman Mike Hanson of Cody, Wyoming, have adapted to the bear and the land restrictions that go with it. "I feel strongly that there is a place for the bear here," says Hanson, owner of the Cody Lumber Company, "but there is also a place for man here too."

Hanson's lumber mill specializes in high-quality wood for paneling, trim and rustic siding. Hanson needs about 5 million board feet of timber a year to keep the thirty-six full-time and twenty part-time mill workers on the job. Most of his wood comes out of prime grizzly habitat, and he has adjusted his operations to the strict regulations put in place to protect the bears. He uses only the most primitive of roads to get into his logging areas and when he's done he abandons them. The logging area is closed in the fall, so Hanson had to buy additional equipment to get the wood out faster, before the fall closing. Now, loggers aren't in the woods disturbing the bears when the bears use the area. Hanson also had to stretch out his cutting plans, so he cuts in five years to get what normally he would have cut in three years. Moreover, he selectively cuts the lodgepole pine instead of clearcutting, as many of his competitors do. "I know when a push comes to a shove the bear would win over me," says Hanson. "The United States wouldn't miss Cody Lumber Company, but it would miss the grizzly bear."

A push came to a shove for Idaho sheep ranchers Sam and Jim Davis. The Monteview, Idaho, brothers and their family had been running sheep on the west slope of the Teton Mountains since the turn of the century. But that ended in 1988.

In the late 1970s, the area where the Davises grazed their sheep had become infamous among grizzly researchers and managers as a place where grizzlies entered but rarely left. Official figures showed that seven bears were killed by sheepherders in that area in 1978 and 1979. But instead of stopping the grazing, U.S. Forest Service officials sent biologists in with the sheepherders to keep an eye on them. So while the herders watched the sheep, the biologists guarded the bears. This immediately improved bear survival in the area. But the uneasy peace didn't last long.

44   In 1987, a bear got to the Davises sheep, and according to the guidelines on their grazing allotment, the sheep had to be moved. When the bear returned again in 1988, the Forest Service offered a different allotment to the Davises, but they turned it down. Like many others in the western sheep business, they decided it wasn't profitable and they quit.

That was not the end of the Davises for the Forest Service. In 1989, Senator James McClure (R-Idaho) slipped an amendment into an appropriations bill that gave the Davises $85,000 with no strings attached as compensation for sheep losses. McClure, who opposed all but the most cursory restrictions on commodity use for the protection of endangered species, wanted to make a point. He wanted to set a precedent that if a rancher was put out of business by the Endangered Species Act, then he should be compensated by the federal government. The problem was that there was little or no evidence that the grizzly was responsible for the kind of losses authorized by McClure's amendment. "We didn't think that amount could be justified," says Bryant Christiansen, a Forest Service range manager.

Sheep ranchers are about the only group that has been significantly limited by grizzly protection. Virtually all of the sheep ranchers in the area immediately surrounding Yellowstone National Park in Situation 1 habitat have been forced to move or retire. Forest Service officials doubt they will ever be allowed to return. But low lamb and wool prices, not grizzly protection, are having the greatest detrimental effect on sheep ranching. Many observers doubt there will be many sheep ranchers left in the mountains around Yellowstone by the turn of the century, with or without bear restrictions.

Still, some sheep ranchers have learned to live with grizzlies. Dick Egbert, an eighty-four-year-old former Idaho state senator has been sheep ranching just south of the Davises' allotment for fifty years. In 1990, federal trappers captured a three-year-old grizzly on Egbert's allotment north of Tetonia. Targhee National Forest and National Park Service (NPS) workers moved the bear to Yellowstone National Park. The sheep killings continued until trappers caught a black bear. There was no evidence the grizzly had killed the sheep, and under the guidelines the bear shouldn't have been moved, according to Idaho Department of Fish and Game officials. As long as the feds don't force him out, Egbert says he hopes to stay. Stan Boyd, Idaho Woolgrowers Association executive director, says the other remaining sheep ranchers in the region want assurances that they will not be forced off their range. "They are willing to live with the bear. Predatory losses are a part of the industry," says Boyd. "We simply want to utilize that range, and when there is a problem, we want immediate and effective action." 45

Federal land conflicts are only part of the challenge. About 1 percent of the land in the Greater Yellowstone Ecosystem is private, yet 80 percent of human-bear conflicts occur on that land, says Steve Mealey, former Shoshone National Forest supervisor and once a grizzly researcher with Knight. Developments on private land keep bears from using all of the available habitat effectively. But worse, the food and garbage usually found on developed lands affect bears the way heroin does an addict. Once they get a taste they want more and more. Garbage dumps in Yellowstone's surrounding communities, such as Montana's West Yellowstone, Cooke City and Gardiner, have attracted bears to these easy, high-energy food sources, thus habituating them to humans and often leading to the bears' deaths or removal to zoos. Although the dumps in these communities have been cleaned up, the problems continue at private residences. These and other deadly areas for bears—areas of bear-human confrontation—throughout the ecosystem become what congressional researchers have called "black holes."

Black holes in space are hypothetical, invisible, collapsed stars with a small diameter and an intense gravitational field that will not allow light and matter that come into their vicinity to escape. The term as used in relation to grizzlies appeared in a 1986 Congressional Research Service report on the Greater Yellowstone Ecosystem. In the Yellowstone ecosystem, black holes are areas where bears enter and often do not escape. The number of black holes appears to have declined in the 1980s as sheep grazing decreased, clean hunting camps were required and enforcement improved. But several remain. Bears continue to clash with humans around Lake Yellowstone. The Church Universal and Triumphant, a New Age religious sect, has carrot fields and orchards just north of Yellowstone in the heart of grizzly habitat. A new mine planned for the mountains above Cooke City could bring 300 workers into the area, increasing private residences and garbage in a regular bear migration route. "There still are too many places for bears to get into trouble," says Louisa Willcox, program manager for the Greater Yellowstone Coalition, a group of regional and national environmentalists dedicated to preserving the Greater Yellowstone Ecosystem.

Managers point to better education and management programs that are minimizing the amount of food and garbage available to bears in and around the park. National forest administrators are providing hunters with bear-proof containers and meat winches to help them keep their camps from attracting bears. Yellowstone even closes some areas to day use to allow bears to live undisturbed. Yet bear-human conflicts will never be completely eliminated. Yellowstone managers in particular remain disturbingly complacent about visitor pressures that already are overwhelming National Park Service personnel. Yellowstone rangers are overextended, often responding only when bear-human encounters take place. The wide range of park activities, from road construction to campground management and garbage patrol, has become more than current ranger staffs can handle. The Clinton administration has proposed increasing the Park Service budget, but the steady increase in visitors makes bear management have to compete with many other priorities.

For the bear in Yellowstone to be deemed officially recovered and then delisted, three goals must be met. First, biologists must sight fifteen different sows with cubs each year for six years. Second, in the eighteen bear management units that divide the Yellowstone recovery zone, sixteen must be occupied with females with cubs each year over a six-year period. Third, mortality must not exceed 4 percent of the population estimate, which is based on a three-year average of females with cubs sighted.

The number of sightings of females with cubs plays an obviously important role in determining if Yellowstone recovery has taken place. This is disturbing to many biologists, including Derek Craighead, John's son and associate, and John Mitchell, a biologist with the Craighead Wildlands Research Institute in Missoula, because it doesn't correlate with the historical Craighead population data in Yellowstone and depends too much on the skill of the observers.

The traditional method used by biologists to estimate wildlife populations is to mark or radio-collar a certain number of subject animals, monitor sightings of both marked and unmarked animals and then determine the population based on the ratio between marked and unmarked animals observed. Knight takes a different approach. He estimates the percentage of females with cubs in the population and then adds together the number of females with cubs seen in the last three years and subtracts any known mortalities. He then divides the number of females with cubs seen by the percentage of females with cubs estimated to be in the population. In 1992 his estimate was 228. Mitchell says Knight's method is too subjective, since it depends on observers distinguishing between different unmarked females. The method also depends on the accuracy of the estimate of the percentage of females with cubs in the population. "It's not good science," Mitchell says. What bothers Mitchell and Craighead is that if Knight's estimate of the percentage of females with cubs is wrong, or if a few sows are misidentified, the actual number could be far different.

Knight's population estimation formula would not be considered a problem by other scientists if it were used only for the purpose he originally intended—to determine the minimum population size. Perhaps the most disturbing problem as far as the future of grizzly bears is concerned is that land managers and even some scientists in the region are using Knight's figures to determine the trend of the population. In other words, since Knight's estimates of minimum population size appear to show a growing population, managers are assuming that, in fact, the population is growing, even though the population could actually be static or dropping.

The variability problem is intensified in the recovery plan proposal because it uses the minimum population estimate to determine how many bears can be killed annually without hurting the population. While the plan takes into account the possible variables by using a conservative allowable mortality figure—4 percent instead of the usually accepted figure of 6 percent—the potential for wide variability in the minimum population estimate using Knight's formula could lead officials to approve a much higher allowable mortality. If the estimate of the minimum population size is way off, enough bears could be allowed to die in one year to send the population into decline, while managers believe it is actually rising.

Environmentalists are especially concerned about the way the minimum population size estimate is used in the Northern Continental Divide recovery area. There, biologists figure into their sightings of females with cubs observations by untrained forest personnel, loggers and others. They say that people who would like to see the bear delisted and the restrictions on land use relaxed could turn in false reports and skew the information.

48   John Weaver, a respected Missoula bear and wolf researcher and a former Forest Service grizzly bear habitat coordinator, supports using sows with cubs as a population index. He says he has confidence in Knight's methods and the people monitoring bears in Yellowstone. "When people like Steve French say they're confident they can make the delineation, I'm willing to accept that," Weaver says.

Knight defended his estimate at the 1992 Missoula conference, acknowledging that the data is subjective, "but I feel good about it so far."

That kind of offhand, unsupported remark gets Knight in trouble with his scientific colleagues. In the world of bear biology few find room for flip comments in debates over lifelong studies.

"I think there's a healthy bit of skepticism among scientists on the status of the Yellowstone grizzly," Weaver says. "I think it's still healthy to be skeptical."

While most managers and researchers are concentrating on recovery, plans for what happens once it is achieved are being discussed. And there are very different perceptions of what kind of activities will be permitted in occupied grizzly habitat once the bear is removed from the Endangered Species Act's threatened list. Some people believe that once the bear is delisted, such activities as mining, ranching, logging and oil drilling will be allowed to resume at the same levels as before listing. But most scientists say management after delisting will look a lot like management under the Endangered Species Act.

There could be minor changes. Once the bear is delisted, Wyoming and Montana hunters once again may get to stalk grizzly bears near Yellowstone. Other changes may not be so minor. The Bridger-Teton National Forest management plan would allow oil exploration companies to explore for oil and gas in Situation 1 grizzly habitat, under strict conditions, even before recovery.

One place where management conditions should not change much is in the nation's first national park, Yellowstone, established in 1872. Unfortunately, even in the park, the needs of the grizzly often have been relegated behind the desires of visitors. Yellowstone National Park officials promised to phase out visitor facilities at Fishing Bridge, a developed area north of Yellowstone Lake in a key migration route for grizzlies. Despite overwhelming scientific evidence that visitor use there displaces countless grizzlies and results in the deaths of others, the Park Service has dragged its feet on closing dormitories, a service station and other visitor facilities. Even if the Park Service does follow through on its promises, a recreational park and visitor center would remain, a monument to the most preservationist agency in the government's unwillingness to make room for Yellowstone's wildest inhabitant. Critics argue that the entire development and Grant Village, a developed area located on the south shore of Yellowstone Lake, also should be removed because both areas keep bears from using all available habitat.

At Lake Hotel, where several creeks fill with spawning cutthroats in the spring, bears are moved to protect visitors. Despite researcher Mattson's clear recommendation to open the hotel later to allow this important feeding activity, park officials stand firmly on the side of visitor service over grizzly bear protection. The same practice takes place on the road overlooking Antelope Creek south of Tower Junction in the northeast section of the park, where Bear 59 regularly hunted elk calves. In the summer of 1990 a crowd of more than 250 people regularly surrounded a young male bear feeding near the road. He eventually was moved to protect both himself and the public, but the precious feeding area is now practically off-limits to other bears, since the Park Service moves bears that conflict with humans.

The Park Service still must deal with the problem of these "neutral" bears that allow humans to approach closely. Such bears often are harassed into a violent response, ending in injury or death for humans and death for the bear. "There is a segment of the population that wants to get closer," says Steve Frye, a former Yellowstone ranger who in 1992 served as chief ranger at Glacier National Park. Seeing bears is what many people expect in Yellowstone, and the park is considering alternatives for allowing observation of bears. But it is moving cautiously. "Is it worth the potential loss of one bear to get this amount of people converted to our program?" he asks.

Cindy Sorg Swanson, a Forest Service researcher from Missoula, suggests that closer contact could actually help the effort to preserve grizzly bears. "I believe under a highly controlled situation, bear viewing could be offered in the recovery area," says Swanson. She says allowing limited viewing opportunities similar to those offered along Alaskan salmon-spawning streams would increase the economic value of bears and strengthen arguments for their preservation.

Kerry Gunther, a Yellowstone management biologist, says increasing viewing opportunities inside the park would be harder for managers than at the relatively isolated Alaskan sites, since nearly 3 million people visit Yellowstone. Steve French suggests that the park install better parking areas along Antelope Creek. Bears usually are seen at long, safe distances there, and perhaps spotter scopes and interpretive programs could be used. "Even when they see a bear at a distance it makes a person's trip," French notes. "They appreciate it more."

Gunther says Yellowstone's program to reduce bear-human conflicts and to force bears to forage for natural foods has been a success. Property damage by bears has dropped from an average 138 incidents annually from 1931 to 1969 to an average of 17 from 1984 to 1991. Annual injuries to humans by bears has dropped from an average of 48 per year from 1971 to 1984 to an average of only 1 per year from 1984 to 1991. "Most of these injuries were caused in the backcountry, and I think it would be hard to reduce this any further," Gunther says.

From 1971 to 1974, rangers trapped an average of thirty-seven bears per year and moved them to the backcountry. From 1973 to 1983 the average dropped to six per year. Since 1983 the average has dropped only slightly to five, Gunther says. However, most of the bears moved out of developments today—73 percent—were feeding on natural foods, compared with 54 percent that were feeding on human foods from 1975 to 1983. "I think this shows we have reduced our nuisance bear incidents about as far as we can unless we go to new, innovative management," says Gunther. He believes that as the bear population and the number of visitors increase, keeping bears and humans apart will be harder. "The problem of having habituated bears in our developments is only going to get worse," he says.

Managers try to control neutral bears through aversion training, in which problem bears are shot with rubber bullets when they enter areas frequently used by humans. The practice is not universally supported by researchers and managers. Proponents say it can be used as an alternative to hunting to keep bears from becoming comfortable around humans. Opponents say it just makes bears mean.

All management will be fruitless if Yellowstone's bears continue to be isolated from the rest of the West's remaining grizzly populations. Despite its apparent short-term health, the Yellowstone grizzly population is still vulnerable to genetic inbreeding that weakens the species. Catastrophic events, such as the 1988 Yellowstone fires, also could destroy the habitat base in Yellowstone, forcing the relatively small number of bears into extinction. The long-term effects of the fires on bears remains unknown. Scientists do know that precious whitebark pine trees were burned, removing part of the important fall food source from several areas in the ecosystem for more than a century.

51

The Interagency Grizzly Bear Committee took its first positive step toward linking Yellowstone's grizzly population to others in the northern Rockies in 1991, when it began efforts to recover the grizzly population in the Selway-Bitterroot Wilderness in central Idaho. The recovery of bears in the still-wild area is a necessary stepping-stone to the long-term recovery of bears in Yellowstone, 240 miles away, and the rest of the northern Rockies.

The last known grizzly was killed in the Selway-Bitterroots in 1956. Despite the use of remote cameras in the summers of 1990 and 1991 and quick follow-ups to reported sightings, biologists have been unable to confirm whether that desolate country still holds a remnant population or contains transient bears from surrounding areas.

In addition to beginning efforts to recover bears in the Selway-Bitterroot, the interagency committee started a program to study the areas between the various recovery areas. In these linkage zones biologists hope to preserve enough habitat to provide travel corridors allowing movement between the last of the large sections of bear habitat. Wayne Melquist of the Idaho Department of Fish and Game says the links between Canada and Yellowstone may remain and should be protected through management. That means limiting road building and keeping the bears and the public apart. If bears can't interbreed up and down the northern Rockies, then each smaller population will have to fend for itself. Just as generals are warned not to divide their armies in the face of superior force, endangered animal populations must stay together biologically to survive the threats to their existence.

Even with enough suitable habitat, the grizzly's future depends on factors such as distribution, age and genetic characteristics. There must be enough bears to prevent inbreeding—the breeding of family members that leads to birth defects—or a deteriorating genetic base, which slowly removes characteristics necessary for adapting to changes in environment.

In isolated populations, the number of animals that participate in breeding is one of the keys to survival. For the population to be considered viable—capable of living, growing and developing as an independent unit—requires a certain number of bears. Viability is calculated on the basis of probability of survival over a certain period of time. Biologists take into consideration the amount of range necessary for each individual bear to survive, the food available, the mortality predicted and such factors as genetics and random catastrophies to figure the minimum number of bears necessary for them to survive until a specific time in the future. For Yellowstone's grizzlies, managers say there is a 95 percent probability bears will survive 100 years.

Conservation biologists distinguish between the population of bears and the breeding population, since the entire population does not participate in breeding. Grizzly sows do not begin breeding until they are three years old and do not breed every year. Many males also don't participate because older, stronger males prevent them from breeding by keeping them away from females.

Biologist Mark Shaffer estimates that a breeding population of at least fifty grizzlies would have a moderately high probability to survive 114 years. His model for determining the minimum viability of a bear population is the most generally accepted. Yellowstone's population of at least 200 bears provides the cushion to ensure at least fifty in the breeding population, Knight says. Mattson says the probability is high that grizzlies won't go extinct in Yellowstone in the next thirty years, but after that the odds drop.

Mattson and Reid say the major problem facing the long-term health of the grizzly is a predicted change in the climate of the region. Harold Picton, a biologist at Montana State University, says the so-called greenhouse effect, or the warming of the earth because of increased carbon dioxide and other gases, is expected to make the already dry Yellowstone range even drier. "It is the consensus of the climate experts that we will be having a major warming of the climate over the next 100 years," says Picton. "At this time we are not taking into account these climatic changes in grizzly management or forest regeneration plans."

Mattson worries about the army cutworm moth, the whitebark pine cone and elk carcasses—three of the grizzly's key food sources, which are found in high altitudes. If global warming significantly changes Yellowstone's high country, those sources could diminish. Elk may move to lower elevations where bears will be more apt to come in contact with humans. Additionally, a combination of climatic factors, such as more fires or long droughts, may transform Yellowstone's vegetative base, on which grizzlies and their prey survive.

The grizzly's range stretched south into Mexico as late as the 1930s,   53 providing bears with a wide variety of range, climates and habitat in which to thrive and adapt. It's likely that bears on the former southern fringes of that range had developed specialized characteristics to live in the Southwest, an environment very different from the northern part of their range. But today, the grizzly bear's southern edge is in the Greater Yellowstone Ecosystem. In an effort to extend that southern edge, the Interagency Grizzly Bear Committee is considering reintroducing bears to Colorado, although such a plan is of low priority, following other, more promising reintroduction programs under consideration in places such as the Bitterroots and the North Cascades in Washington. If one considers what would happen if climatic changes or other events made the northern end of the bear's range less desirable, extending today's southern fringe would become even more important than it seems today.

Yellowstone's grizzly baby boom of the late 1980s has had more of an effect than simply making national park and national forest managers more complacent. As the very territorial grizzly bears filled the best habitat in the park and in desolate country, more and more bears have been pushed to the marginal range on the fringes of the ecosystem. As bears wander out of national parks and wilderness areas looking for their own turf, conflicts are bound to increase. "We're going to see more of this as the population recovers," says Forest Service range manager Bryant Christensen.

While bears seem to be expanding their range south and east, they are apparently avoiding the fringes outside Yellowstone's western border. That area, which includes the heavily logged Island Park area in Idaho and Montana's Madison Range to the north, was initially included in the recovery zone because it was considered occupied by bears. But a decade of overcutting, heavy recreational use and development has made the area less inviting to bears. The interagency committee has initiated a study of the Island Park, Idaho, area to see if it can, in fact, support bears. Meanwhile there is clear evidence that grizzlies are expanding their range into the Wind River Mountains to the southeast of Yellowstone.

Even though the bears are expanding their range, some managers are hesitant to expand the boundaries of the recovery area to include the areas newly occupied by the bears. "We recognize that we have bears outside the recovery zone," says Wyoming Game, Fish and Park's John Talbott. "Regardless, our emphasis as managers is on bears inside the recovery zone."

Though researchers Mattson and Reid question whether the current recovery area is large enough to support enough grizzlies for full recovery, expanding the recovery zone is not a favored option for the state and federal officials on the Yellowstone subcommittee of the Interagency Grizzly Bear Committee. "If there are to be adjustments in the line, they will be based on information that more habitat is needed to protect the viability of the population," Chris Servheen says.

Politics clearly will decide whether the bear will get more space. Yellowstone superintendent Robert Barbee has warned that "willy-nilly" expansion of the recovery zone will damage the credibility of the committee's agencies with ranchers and others living near the park. He worries that changing the boundaries for bears will make those already skeptical groups more opposed to wolf reintroduction, one of his pet projects. Barbee's sensitivity arises from years of fighting against the powerful political forces of ranching and other economic interest groups.

If bears aren't legislated a larger sanctuary, then ensuring their protection inside the current recovery zone will continue to be the major issue. The bottom line is that we humans need to learn to share the land with what is often a dangerous and uncompromising neighbor. "Sometimes you just can't negotiate and protect the bear," says Steve Mealey.

The grizzly's best hope in Yellowstone and the rest of the northern Rockies lies in protection of its habitat and better knowledge of what it needs and how it adapts. The grizzly bear's uncompromising nature, its need for privacy and its low birth rate all work against its long-term survival. Yet the bear's capacity to learn new hunting techniques and other skills, and its ability to pass such learning on to new generations, give some hope that it can adapt if humans are willing to give it places to do so.

Richard Knight says that he and his research team have seen the bear come a long way since they began their work in 1973. "There were an awful lot of dumb bears then, like we were dumb," he says. "We've kind of learned together."

In the last two decades the educational process has not been limited to bears and biologists. As hard as the lessons have been, sheep ranchers have learned to stop shooting grizzly bears, and foresters have learned to manage forests so that grizzly bears can live where loggers 55 have cut. Yet if the grizzly bear is to survive in the northern Rockies, the learning curve must continue upward faster than the bear population itself. Americans must pass on to the next generation the knowledge that there are limits to how much intrusion bears can stand into their wilderness home. Humans cannot continue to take valuable bear habitat without giving something back in return.

3. Troubled Waters

AT FIRST glance, the endangered bald eagles of the Greater Yellowstone Ecosystem don't appear to have any relationship to the potato crop growing on the wide Snake River Plain to the west in Idaho. For that matter, they don't seem to have much in common with the trumpeter swans that winter in Idaho's Harriman State Park, the whooping cranes that spend the summer at Grays Lake National Wildlife Refuge east of Idaho Falls, Idaho, five rare snail species near Hagerman, Idaho, the salmon that run up the Columbia, Snake and Salmon rivers, the fishermen who depend on the salmon or the industries that depend on cheap power from hydroelectric dams to fuel the Pacific Northwest economy.

Their common link is the need for water, often the same water. In the West, water's scarcity almost defines the region. It is a precious substance on which all life and much of the economy depends. Who gets how much decides the winners and losers in the Pacific Northwest economy and in nature. Without water, the cranes wouldn't have a place to nest and potato farmers couldn't grow a crop. The Henry's Fork of the Snake River, one of America's premier trout streams, would freeze over and eliminate the trumpeter swan's winter home. The nationally famous trout fishery of the South Fork of the Snake River would dry up, depriving the eagles of a key food source. Five snails that have outlived most of their Pleistocene cousins would disappear. The salmon—which carry economic, aesthetic and even spiritual value—would never return to the mountain streams of Idaho. And who gets how much water could change dramatically now that Snake River salmon have been protected under the federal Endangered Species Act.

Water in the Pacific Northwest has been controlled historically by agricultural interests and hydroelectric power generators. The two groups have frequently battled each other for control, but until now nothing has seriously challenged their right to do what they please with the region's water. The power of the Endangered Species Act, unleashed when Snake River sockeye salmon were listed as an endangered species in 1991, provides the legal muscle to wrestle the hands of farmers and electricity generators from the tap of the Northwest's faucet.

Tradition plays a large role in the management of water. Idaho water law, like that throughout most of the West, is based on the doctrine of prior appropriation: whoever used it first has the first right to it. Started in the mining camps of California, this primary rule of western water law has survived for more than a century and ensured that irrigators who settled Idaho's Snake River Plain beginning in the 1870s would control most of the waters running out of the mountains of Yellowstone and Grand Teton national parks as well as much of eastern Idaho and western Wyoming.

Since the Snake River Plain is a desert, most of its precipitation comes in the form of snow along the higher elevations. More than sixty inches of precipitation in the form of snow falls in some areas. Today, the Idaho Department of Water Resources (IDWR) says more than 30,000 farmers irrigate about 3.8 million acres of land in Idaho's Snake River Valley, which covers most of southern Idaho. About a million of those acres are in the federal Bureau of Reclamation's Minidoka Project, a series of dams and ditches that supply stored water to farmers from Ashton, Idaho, on the southwest corner of Yellowstone to King's Hill, a hundred miles southeast of Boise. These farmers grow potatoes, hay, wheat, barley, sugar beets, beans and other crops worth about $350 million annually, according to the bureau. Their income provides the economic base for most of southern Idaho, especially its small towns. These farmers hold the rights to virtually all of the natural flow in the Snake River and its tributaries and most of the 4.5 million acre-feet of water stored mostly in four major reservoirs: Jackson Lake, in Jackson Hole, Wyoming; Palisades Lake, on the Wyoming-Idaho border; Island Park, on the Henry's Fork of the Snake River directly west of Yellowstone; and American Falls, west of Pocatello, Idaho. These irrigators have diverted the flow of the entire Snake River at Milner Dam near Burley since early in the century. The river continues to the Columbia only because of the flows of thousands of underground springs that enter the river from the Snake River Plain Aquifer along a forty-mile stretch from Milner west to Hagerman, Idaho. The irrigators who control the water above this point call any water that flows over the small Milner Dam wasted.

Even though the federal government built the dams that created the reservoirs, irrigators paid a portion of the cost and believe that fact gives them special rights to the water beyond even the traditional legal right through prior appropriation. The Bureau of Reclamation actually operates the dams, but the federal agency does so strictly under the direction of the "watermaster" of Water District 1, a state-designated area covering the 300-mile-long Snake River Plain area. The watermaster is chosen by a quasi-governmental group known as the Committee of Nine. The committee represents all of the irrigation districts and canal companies in the Snake River Valley. While it is supposed to be only an advisory group, realistically it has controlled water use in the valley since it was formed in 1919 to end a fight over water between irrigators north of Blackfoot, Idaho, and irrigators south of Blackfoot. Ronald Carlson, the current watermaster, is paid by both the Committee of Nine and the state of Idaho, which he serves also as the eastern Idaho IDWR regional director.

As watermaster, Carlson distributes the water to the various canals according to their priority for natural flows and the storage space they own in the various reservoirs. He tells the Bureau of Reclamation when to release water from the dams based on direction from the Committee of Nine. If irrigators want to increase flows in the river to help fish or wildlife, which at times they are apt to do, he orders it. If they want to dry up the river—which was done by mistake briefly in the spring of 1992—he can order that too. The Bureau of Reclamation has not challenged this authority up till now. In fact, it defended its dam management policies in 1987 when members of Trout Unlimited in Idaho Falls, Idaho, sued to force the agency to increase winter releases from Palisades Dam to preserve cutthroat trout populations. The U.S. 9th Circuit Court of Appeals in San Francisco upheld the decision of a lower court that said the agency was not required to conduct an environmental impact statement analysis even if it reduced flows from a dam so low that fish were trapped by the hundreds of thousands in side channels and died.

Legally, the state of Idaho owns the water and appropriates it under state law. The federal government holds the rights to water reserved for federal lands and federal uses. But the old system of prior appropria-   59
tion is coming under increasing court challenge. Lawyers, environmentalists and other users are arguing that the state and federal governments should allocate water to protect the public's interests. This so-called public trust doctrine is based on the idea that public water uses, such as instream flows for fish, wildlife and recreation, must be protected. It is a direct challenge to the common western belief that irrigation is the highest and best use of water.

The Bureau of Reclamation could be forced, either by the National Marine Fisheries Service (NMFS) or by court order, to release water from its reservoirs in the spring under the mandate of the Endangered Species Act. A judge may order that irrigators be compensated for their losses, but the act could be the impetus that forces a revision of water laws in Idaho to allow leasing of water to restore salmon. That possibility sends fear through the irrigation community.

Potatoes are Idaho's most famous irrigated crop. The state produces nearly a quarter of the nation's supply, about 100 million hundred-pound bags annually, worth about $300 million depending on prices. Yet cattle and wheat each still return more to Idaho's economy than potatoes in all but banner years. Idaho's potato farmers are not the only ones with a stake in the outcome of the salmon water fight. Western and northern Idaho farmers and Oregon and Washington farmers also depend on irrigation water from the Snake and Columbia rivers. Overall, irrigation farming accounts for $5 billion in income in the Pacific Northwest.

The Pacific Northwest's food-processing industry is intrinsically dependent on the irrigators and their ability to raise crops. In fact, the food-processing industry provides a good example of the interrelated economic effects that will be caused by efforts to recover salmon. The industry accounts for more than $4.6 billion in sales and employs nearly 60,000 workers, mostly in the region's economically imperiled rural areas. The Pacific Northwest produces more than 50 percent of the nation's frozen vegetables. More than 90 percent of the food processors' products are shipped out of the region. The industry has overcome the high transportation costs to eastern U.S. and foreign markets with its access to water, abundant raw food crops and inexpensive electrical power. Changes in agriculture or power rates could "jeopardize this delicate economic balance," says David A. Pahl, president of the Northwest Food Processors Association.

60

Before farmers began diverting the Snake River and its tributaries in the 1870s, the Shoshone and Bannock tribes lived in the Snake River Valley. Tribal water rights are based partly on the fact that the tribes were using the water before white people settled in the area. Historically, the Shoshone and Bannock tribes migrated widely around the Snake River Basin, where they fished, hunted and lived off the bounty provided by the land and waters through the seasons. The Shoshones were Sacajawea's people, who met Lewis and Clark when the explorers came over Lemhi Pass into Idaho in 1805. Even after the Shoshones were moved to the Fort Hall Indian Reservation near Pocatello, Idaho, in 1855 by the Fort Bridger Treaty, the traditional migration continued, with the Salmon River the main fishing ground. "The salmon is a part of the Indian people's life," says Keith Tinno, a Shoshone and tribal business council member. "I remember my grandfather telling me they would travel by horse and buggy or Model T toward the headwaters, the East Fork, the Yankee Fork and other creeks."

The Shoshone-Bannock tribes, thrown together by the federal government, have worked to preserve their natural heritage—water and salmon. An historic water rights agreement between the Shoshone-Bannock tribes and the state, approved by the U.S. Congress and signed by President Bush in 1991, allows the tribe to leave a portion of its water in the Snake River in the spring to help flush migrating juvenile salmon from Idaho back to the ocean. The tribes also filed a petition with the NMFS asking that the Snake River sockeye salmon be protected under the federal Endangered Species Act. Later, environmental groups added petitions to protect Snake River chinook and Columbia River coho. But it was the Indian tribes that forced the issue. Sockeye were listed as an endangered species in November 1991. The Sho-Bans, as they call themselves, broke with Indian tradition by negotiating the water rights agreement rather than taking the case to court. The listing petition also set the tribes apart.

In the 1970s, Indian tribes in Washington, Oregon, Montana and northern Idaho won important court cases reserving their right to salmon on the Columbia River and lower Snake River in Washington and Oregon. The case, *U.S. vs. Oregon,* recognized the rights of tribes downstream to half of the salmon taken in the Pacific Northwest. But the Sho-Bans, who were not parties to the treaties of 1855 and 1856 signed by Oregon, Montana, Washington and northern Idaho tribes and Washington territorial governor Isaac Stevens, were left out. When the states of Washington and Oregon reached an agreement with the downstream tribes to comanage salmon harvests, Idaho and the Sho-Bans were again left out.

The only way the Sho-Bans could protect the salmon that return to their traditional Idaho spawning grounds was to go over the heads of the states and the other tribes to the federal Endangered Species Act.

The Sho-Bans' independent actions have often angered the other Pacific Northwest tribes. In 1992, for instance, the Sho-Bans went to court and won a lawsuit with the Nez Perce tribe of northern Idaho in a dispute regarding the boundaries of traditional fishing grounds. Essentially, however, the Sho-Bans' interests in salmon fishing simply don't coincide with those of the downstream tribes. The biggest difference concerns whether efforts should be focused on preserving wild salmon or producing higher numbers of salmon from hatcheries. The Sho-Bans want everything possible done to preserve wild stocks, while the downstream tribes want to enhance the wild stocks by planting hatchery fish in unused wild habitat. The Sho-Bans and many environmentalists say such outplantings will dilute the genetic base of wild salmon and weaken those stocks that have survived. At meetings with other tribal leaders and the states, Sho-Ban leaders have often been noticeably shunned by other tribal leaders. In the case of the Nez Perce it's a feud that goes back before the treaties. This behavior doesn't deter Keith Tinno. For him, saving salmon is not simply an effort to bring back fish to Sho-Ban tables. It is a spiritual quest. "The salmon is to me the spiritual connection," Tinno says. "I don't want to lose the wild salmon."

In recent years, tribal fishermen have continued to exercise their treaty rights to fish for spawning salmon returning to the tributaries of the Salmon River in central Idaho despite the low numbers of fish left. This has come often at the objection of the Idaho Department of Fish and Game (IDFG) and Idaho fishermen. Sometimes the catch has been limited to dozens instead of the thousands taken by downstream tribes. Tinno says fishing limits are controversial in tribal politics. The Shoshone and Bannock people remain poor, and unemployment is high. Wild game and fish still are important food sources, in addition to being the spiritual and cultural links with their ancestors. Limiting harvests, especially to only a symbolic catch, as has been done often during the last decade, is as popular among tribal constituents as cutting entitlements is to other American voters. In fact, several tribal leaders have lost their jobs for championing limits. But Tinno says even though he supports the limits, he strongly believes in the need to continue the fishing tradition. The loss of traditional salmon fishing among Idaho sportsmen has contributed to the decline of the fishery because many people don't care about what they can't see and feel in their hands. The political pressure to preserve salmon has diminished with the salmon fishing tradition, Tinno says. "But some white people feel the same way we do," Tinno says. "They don't want to lose a part of what their ancestors did."

The Sho-Ban water agreement already broke new ground in Idaho water policy by authorizing the tribes to establish a "water bank" of stored water that allows them to market excess water, both with long- and short-term leases. The agreement permits the tribes to lease the water outside the upper Snake area east of Milner Dam, something that was not allowed previously by irrigators who control the water. Idaho water banks—relatively new themselves—formerly were authorized only to enter into short-term leases to lease water within the specific water districts.

Since the tribal agreement, the Idaho State Legislature authorized a short-term leasing program that even allows irrigators to lease water out-of-state. That water, as well as the Sho-Ban water, will be used to flush juvenile salmon and steelhead downstream through the four hydroelectric dams on the lower Snake that act as barriers to salmon migration.

This leasing program offers irrigators the same opportunity that the Sho-Bans have to make additional money by saving salmon. It's an opportunity few irrigators want to take since they believe that water marketing threatens irrigation farming. That's because the water is now leased to other irrigators at $2.75 an acre-foot, about the cost of transporting the water from reservoir to farm. In a drought, many water-poor farmers depend on this cheap water to bring in their crops.

In 1992, the Bonneville Power Administration (BPA) was offering farmers $20 an acre-foot to leave their water in the Snake River so that it could run through Idaho Power Company's hydroelectric turbines and downriver to help flush young salmon to the sea. The savings in power production, by using hydroelectric generation instead of a coal-fired plant, more than makes up for the higher cost of the water to the BPA. So if farmers begin leasing their water to hydro utilities to help salmon, it will price their neighbors out of the market. This has prevented states such as Idaho from allowing a free market in water sales.

Despite irrigators' fears, the listing of Snake River salmon could make them money rather than cost them, says Ed Chaney of Save Our Wild Salmon in Eagle, Idaho. Instead of taking irrigators' water when it is still needed for crops, federal officials more likely will lease their surplus when available. Or the BPA could find it cost-effective to pay for water conservation measures that benefit both farmers and ratepayers and make more water available for leasing. It could even give Idaho's irrigators more flexibility in managing their crops. Some farmers may decide to lease their water one year rather than risk planting in a poor water year or when crop prices are bad. Or they may lease some water and plant crops, such as hay, that use less water. Over time, Idaho irrigators may make the transition to low-water, low-power irrigation systems and make more money off their water than their crops, Chaney says.

The complexity of managing water ecosystems means that sometimes the issue is not just human needs competing with fish and wildlife needs. It can be fish and wildlife versus fish and wildlife. Following the Snake River system from Yellowstone National Park to the Columbia and the ocean, an observer finds many such natural clashes.

Without additional storage, irrigators say any water that goes downstream to salmon may be at the expense of other fisheries, such as the blue-ribbon cutthroat trout fishery in the South Fork of the Snake River east of Idaho Falls, Idaho. That, in turn, could affect the endangered bald eagle's thriving population in eastern Idaho, since the eagle feeds on cutthroat. That eagle population is the strongest of any in the Greater Yellowstone Ecosystem and all of Idaho. When Snake River flows are decreased to fill the Palisades Reservoir, the river below ices up. When that happens, the eagles can't fish. "Low flows mean significantly less winter feeding habitat for bald eagles," says U.S. Bureau of Land Management (BLM) biologist Russ McFarling of Idaho Falls.

Leasing water downriver to help salmon also could threaten the water supply available to keep the Henry's Fork from freezing at Harriman State Park in Island Park, Idaho, and killing Rocky Mountain trumpeter swans. In 1989, at the peak of a drought, more than fifty swans died when flows on the Henry's Fork were cut to the minimum. Subzero temperatures froze the surface of the entire river, cutting off the supply of aquatic plants that the swans depend upon as winter food. They simply starved to death.

Years of manipulation of water and wildlife populations by humans has made the supply of both dependent on humans. Farmers and whooping cranes both depend in part on water management decisions for their survival. The Bureau of Indian Affairs controls flows from the 18,500-acre Grays Lake National Wildlife Refuge east of Idaho Falls, Idaho. The water is used by downstream irrigators to grow potatoes, wheat and barley. It also provides protection for whooping cranes. When the water level in the Grays Lake marsh drops, crane nests, eggs and young birds are vulnerable to predation by coyotes and foxes. Mortality was particularly high in the late 1980s—a period of below-normal precipitation that saw a high demand for irrigation water.

The whooping crane, an international symbol of the plight of endangered wildlife, was introduced to Grays Lake to try to establish another wild flock besides the world's only remaining flock, which spends each winter in Texas. So far, the attempt has failed, and drought has been the primary reason. The flock, which peaked at thirty-seven cranes in the northern Rockies in 1985, had plummeted to thirteen in 1990. Only eight whoopers spent the summer at Grays Lake in 1990, down from twenty-three in 1983. Water management is the limiting factor for cranes at Grays Lake.

Meanwhile, another water dispute has been brewing downstream in an area known as the middle Snake. The middle Snake, starting below the American Falls Reservoir and running west to near Boise, Idaho, is an aquatic ecosystem out of kilter. Erosion from a century of irrigation ditch drainage has covered the river bottom with sediment. Fish farms have added additional nutrients to the river, already diminished by the diversions for irrigation. The result has been serious water quality degradation.

The middle Snake dispute pits fish farmers and power companies against four snails and a limpet. It wouldn't be a fair fight except for the Endangered Species Act. The five mollusks are believed by biologists to live only in small areas along a forty-mile stretch of the Thousand Springs area of the Snake River. In January 1993, the U.S. Fish and Wildlife Service (FWS) listed the the Bliss Rapids snail, the Utah valvata snail, the Snake River physa snail, the Idaho spring snail and the Banbury Springs limpet—a mollusk with a cone-shaped shell and a thick, fleshy foot—as endangered species. They are only found in the 65 cool, clear, free-flowing waters of the Snake River or large adjacent springs in the Hagerman, Idaho, area. The snails are the remnants of what was once a diverse population of ninety mollusk species that lived in Lake Idaho during the Pleistocene epoch, roughly 12,000 years ago. The Banbury Springs limpet, also known as the lanyx, is believed to be the only representative of its genus left in the state.

Earl Hardy, a Boise businessman, has been trying for twenty years to build a fish farm using water diverted from Box Canyon Creek near Hagerman. He has run into stiff opposition and red tape with twelve different agencies and court challenges from several environmental groups. Even without the snails having been listed as endangered species, Hardy would have a hard time getting approval for his project, since it will add to the already high pollution levels along the middle Snake. The listing recommendation for the Banbury springs limpet creates an additional roadblock.

Fish farming is very important business in the Magic Valley, the farm country surrounding the middle Snake. Almost 90 percent of the nation's rainbow trout sold in stores and restaurants comes from fish farms in the Thousand Springs area. Also, the Thousand Springs stretch of the Snake River is a wondrous ecosystem, where the spring water provides life for a host of aquatic species, including huge white sturgeon. The snails are among the most sensitive of the species that depend on the system and therefore are the best indicators of its health.

Hardy's plan and several other projects threatening the Thousand Springs ecosystem are what prompted the FWS to protect the rare mollusks. The major threats are two hydroelectric projects: the A. J. Wiley project, proposed above Bliss by Idaho Power and the city of Idaho Falls, and the smaller Dike hydro project below Bliss. The Wiley project, an 86-megawatt hydroelectric dam, has already run into stiff opposition from environmentalists because its backwaters would flood the last free-flowing stretches of the middle Snake. The mollusk listing under the Endangered Species Act could spell the end of any hydroelectric development in the area, because the snails need flowing water to survive. Hydro developers continue to comb the basin for other sites where the mollusks live; so far, their findings have been inconclusive. The concern over protecting the mollusks underscores the important link between water and the life in it. Aquatic ecosystems are the most sensitive to disturbance. Just about any activity that takes place on the land surrounding them eventually affects the quantity or quality of the water. Mollusks, like salmon, are the indicator species for water ecosystems. They tell us how healthy the natural water ecosystems are. If the mollusks can't survive, then a whole group of species also are imperiled.

The five Snake River mollusks also need flows running from the Thousand Springs that restore the river below Milner Dam. These springs run out of the Snake River Plain Aquifer, which underlies the Snake River Plain from Ashton to King's Hill. Since irrigation started in the late 1800s, the flows from the springs actually have increased, due to recharge of the aquifer from irrigation. Chuck Lobdell, FWS Idaho director, says the mollusks may preserve the practice of irrigation along with the middle Snake ecosystem. If land on the Snake River Plain was taken out of production and irrigation stopped, the flows at Thousand Springs could drop, sending the mollusks into extinction. Lobdell's analysis is not universally shared, yet it demonstrates how complicated changing the current ecosystem to balance both civilization and the natural systems will be in the future.

Another concern in the middle Snake area is the threat of California, Nevada or some other Southwest urban center "stealing" Pacific Northwest water. This is one of the great paranoias of water users in the region, as it is along the middle Snake that California politicians have proposed grandiose diversion schemes. Pacific Northwest political leaders can always get a rise out of voters, particularly in rural areas, by promising to prevent the region's water from getting into the greedy hands of Los Angeles or some other city. Northwest congressmen have successfully put in place a federal law that prohibits even studying such a proposal. Still, because of the power of the California congressional delegation, such a diversion remains a slim possibility. Listing of salmon in the Columbia River, while a potential threat to Idaho water rights holders, also could protect the Pacific Northwest from water diversions to Nevada, California and Arizona. Any diversion on the scale needed to provide water for California would threaten salmon survival as well as the middle Snake ecosystem.

The middle Snake also was the edge of the historic range of salmon in the Snake River. Salmon used to migrate up the Snake River all the way east to Shoshone Falls, a 212-foot-high, 1000-foot-wide falls near

Twin Falls, Idaho, that kept the fish from swimming perhaps all the way to Yellowstone. Early in the century, below Shoshone Falls, dams built to generate electricity to the growing state of Idaho began eating 67 away at salmon habitat on the Snake River. In the 1950s and 1960s, Idaho Power Company built a series of three dams in the Hells Canyon area along the Idaho-Oregon border northwest of Boise. These dams did not provide effective fish ladders to allow adult salmon to swim upstream, and the Brownlee Reservoir, created by the dams, was too long and its current too slow to allow salmon passage. This closed off thousands of miles of salmon habitat in the Snake River and in such important tributaries as the Boise, Weiser and Payette rivers. But since other dams and poor land practices had already blocked off most of these rivers to salmon, little was done when the Hells Canyon dams proved to end salmon migration upriver.

Below Hells Canyon, eight more dams on the Snake and Columbia rivers operated by the U.S. Army Corps of Engineers and the BPA stand between Idaho's spawning grounds and the Pacific Ocean. They act as barriers to migration, particularly to juveniles in the spring. The federal government provided the capital to build these dams, which gives residents of the Pacific Northwest electric power at significantly lower rates than the rest of the nation. While not everyone in the Pacific Northwest gets this subsidized power from the BPA, what the federal agency does with power rates affects just about everyone's bills. The BPA supplies half of the Pacific Northwest's power, serving Oregon, Washington, Idaho, western Montana and portions of Wyoming, Nevada, Utah and California. The BPA sells power wholesale to public and private utilities from a system of thirty dams on the Columbia and Snake river systems, two nuclear power plants and coal-fired plants in Washington and Montana.

Modifying dam operations to improve salmon migration would raise power rates and threaten such power-dependent enterprises as the aluminum industry, which pumped nearly $1.9 billion into the Pacific Northwest economy in 1989 and produced 43 percent of the nation's aluminum, according to the Direct Services Industry Association, an aluminum industry lobby group. The Northwest Power Planning Council, a panel appointed by the governors of Idaho, Oregon, Montana and Washington to advise the BPA on power production and conservation issues, predicted rate increases of about 8 percent for salmon recovery costs. Even the most pessimistic insiders now say the worst possible electricity increase attributable to salmon would be from 15 percent to 20 percent. Current rates in the region average about 4 cents per kilowatt hour.

68

The reservoir and lock systems at the dams allow ocean port access as far upriver as Lewiston, Idaho. The Columbia is the second-largest navigation system in the United States after the Mississippi, with $12.2 billion in commodities moved in 1989. Wheat and other grains are the most important commodities, and wheat shippers say they need to be able to ship year-round to compete for markets in Asia and elsewhere. Idaho wheat sales exceed $300 million annually, according to the Idaho Department of Agriculture. Many years, the value of wheat sales exceeds that of potatoes. The richest wheat-growing area is the Palouse Country of northern Idaho, an area of rolling hills with enough rainfall to grow the grain without irrigation. Together, the wheat crops of Montana and Idaho account for a large portion of Snake River shipping.

Irrigators and most of Idaho's various interest groups following the salmon issue, including environmentalists, argue that simply flushing more water downriver won't solve the salmon migration problem. This viewpoint has put them at odds with a variety of interest groups in Oregon, Washington and even parts of northern Idaho that believe that releasing water from Idaho's upper Snake reservoirs in the spring is the way to help juvenile salmon return to the ocean. The reason that Idaho irrigators and environmentalists have become unlikely allies with Idaho governor Cecil Andrus in an alternative plan is simple arithmetic.

The entire flow of the Snake River from Idaho amounts to only about 36 million acre-feet of water annually. If dam operation on the four lower Snake dams in Oregon and Washington is not changed dramatically, fish and wildlife authorities throughout the region agree, water would have to be run through the dams at a rate of 140,000 cubic feet per second continuously for two months to effectively flush the young salmon to the ocean in anywhere near their historic travel time. That would take about 16 million acre-feet of water run through the lower river in a two-month period. Even if all of Idaho's irrigated lands were dried up and all available storage drained, only about 8 million acre-feet could be made available for the fish. "That is not enough water to do the job," says Governor Andrus, who has led the fight against plans to seek upstream water as the sole salmon migration solution. "Idaho is part of the solution, but we are not going to give up our water."

The BPA has proposed increasing the Snake River storage water made available from the current 450,000 acre-feet to as much as 1 million acre-feet. But even if they could get that much water, by Idaho's estimation it will fall short of the amount needed to move the young salmon to the ocean.

Andrus has proposed draining the reservoirs behind the four lower Snake dams substantially—up to forty feet—during the spring and increasing the velocity of the river. If that were done, according to Andrus, less water would be needed to flush the smolts downstream. The idea behind Andrus's bold plan is to increase the velocity of the river to replicate the speed of the flows that would result if the current system had 140,000 cubic feet per second running in the spring. This is the number scientists have estimated would be optimum for salmon. It would make travel time to the ocean about twelve days. Historically, the trip through the Snake and Columbia rivers took as little as five days, but the reservoirs behind the dams that now catch the spring runoff slowed the trip. Now it takes from seventeen to thirty days.

Unfortunately, the dams and their reservoirs were not engineered to allow the kind of drawdowns proposed by Andrus. The fish passage facilities, especially the fish ladders that allow adults to return to Idaho, won't work at the low flows. Army Corps of Engineers scientists have even predicted that the dams might not hold at the lower levels because the soils around their base might be destabilized when dried out. Roads and railroad beds could be eroded into the river by the faster river flows, the most fearful predict. In a test drawdown conducted by the Corps in 1992, no structural damage was observed in the dams, but several roadbeds cracked and riverside facilities such as docks and grain-loading facilities suffered damages. Andrus believes the costs of damages could be covered by either Northwest ratepayers or taxpayers. He says that to make his plan work, the four Snake River dams will have to be modified. But the cost of modification, according to studies he has commissioned, is affordable. The Army Corps of Engineers, which operates the dams, disagrees. It says changing the dams could cost as much as $4.5 billion, and even then they are not convinced the plan is biologically sound.

70

Lowering the level of the reservoirs as Governor Andrus has proposed could hinder grain shipments by barge from Idaho during the two-month spring migration period. Much of Montana's grain travels by truck to Lewiston, and these shipments carry the added clout of the Big Sky State's two U.S. senators and one congressman if threatened. But Andrus suggests that intermittent pulses of water could be sent downriver to accommodate barge traffic. Or new grain-storage facilities could be built downriver to ensure a steady supply to meet market demands during the period the river is unnavigable.

Farmers in Washington and Oregon claim Governor Andrus's drawdown proposal could take 309,000 acres of land out of production unless pumps and pipes that pull irrigation water out of the reservoirs are improved. "No one is going to dry up hundreds of thousands of acres," Andrus says, pointing out that his plan calls for the federal government to pay the cost of upgrading the pumps.

The National Marine Fisheries Service has formed a committee to study the economic effects of salmon recovery on the region. Norman Whittlesey, an economist from Washington State University who serves on the panel, says drawing down the reservoirs will not threaten

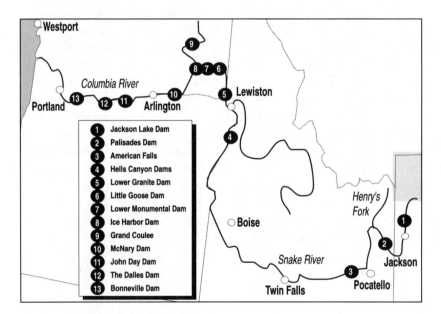

The Key Columbia and Snake River dams.

the economic future of the Pacific Northwest. "My personal feeling is the drawdown problems are very manageable and that the region's economy will survive," Whittlesey said during a seminar in Pullman, Washington, in early 1992. "We have dwelled too much on extremist alarm cries."

Ed Chaney, the salmon advocate from Eagle, Idaho, who drafted Andrus's plan, points out that most of the people complaining about the costs of saving the fish are dependent on heavy subsidies from the federal government. In addition to the massive electricity generation and transmission system operated by the BPA, Army Corps of Engineers and Bureau of Reclamation, the reservoirs that store irrigation water were paid for either totally or partly by the federal government. The lock system that allows barge traffic also was paid for by the federal government. "They've been getting their piece of the action for years," Chaney says. "Now it's time to pay for it by saving the fish."

Once the Snake runs into the Columbia another interest group dominates the debate over fish and water: the Columbia River Indian tribes. At a remote, unmarked boat landing near Arlington, Oregon, in September 1990, a group of Yakima Indians unloaded the day's catch—hundreds of pounds of huge fall chinook salmon. Mike George, a young traditional chief from Goldendale, Washington, placed the fish in large ice coolers as his partners, Mike Olney, Donna Lopez and cousin Sam George, removed the twenty- to forty-pound "fall brights" from the nets they had set overnight in the Columbia River.

Although their tools are different—nylon gill nets and outboard motor boats—the Yakima are following the traditions of their ancestors, who thrived in the region on the bounty available from its rivers. They have a hard time putting a value on the fish they take, even though they get what amounts to a minimum price from the mostly non-Indian fish wholesalers who meet them at the landing. The price is far lower than the fish's selling price in supermarkets.

The Yakimas' ties to the fish are not only commercial, but also traditional and spiritual. The cultures of the Pacific Northwest tribes were based primarily on the availability of fish from the Columbia River. Salmon provided an abundance that allowed the tribes to flourish without having to migrate great distances, as the Great Plains tribes did. Columbia River Indians revere their historic fishing grounds as other tribes revere sacred mountains or as Christians honor Bethlehem or Moslems honor Mecca. Every spring, each tribal group along the river holds a "first salmon" ceremony to celebrate the return of the first fish of the year and the rebirth of the annual cycle of life that connects the Indians with their ancestors, the land and all of life.

The importance of fishing and traditional fishing grounds was demonstrated in 1855 and 1856, when tribes throughout the region gave up 64 million acres in what is now Washington, Oregon, Idaho and Montana in exchange mostly for the retention of the right to fish at "all usual and accustomed fishing places in common with citizens," according to the treaties. These traditional fishing areas, such as Celilo Falls near what is now The Dalles, Oregon, remained important tribal fishing places even as the Columbia River was tamed and slowed by dams.

Wilfred Yallup, a Yakima tribal council member, remembers when The Dalles Dam backwaters covered Celilo Falls in 1957. "They just buried it," Yallup says. "My father said, 'With it they buried my life.' " His father continued to fish, but to him the traditional cycle had been broken. In the 1960s, as traditional tribal values went through a rebirth on and off reservations, a new drive to restore the full extent of treaty fishing rights began.

In 1968, David Sohappy, a Yakima, was laid off from a sawmill and decided to return to the traditional life of his ancestors. Sohappy, who claimed to be a descendant of a Wanapum band chief, settled with his family in a wood house on Cooks Island in the Columbia River, near Cascade Locks, Oregon. He began fishing and along with fourteen members of the Yakima Indian Nation filed suit against Oregon's regulation of off-reservation fishing. Sohappy based his case on the fishing rights reserved in the 1855 and 1856 treaties between Washington, Oregon, Montana and northern Idaho tribes and Washington territorial governor Isaac Stevens. The Yakima, Warm Springs, Umatilla and Nez Perce tribes also sued in a separate case that was to be known as *U.S. vs. Oregon.* In Washington, the Puyallup tribe near Port Angeles filed a similar lawsuit.

The tribes' right to fish in the usual and accustomed places was first recognized in 1969 when U.S. District Court Judge Robert Belloni held in *U.S. vs. Oregon* that the tribes were entitled to a "fair share" of the fish runs and that the state was limited in its ability to regulate "only when reasonable and necessary for conservation." Then in 1974, after a three-year trial, U.S. District Court Judge George Boldt ruled that the

Indian and non-Indian fisheries were entitled to a 50-50 split of the harvestable number of fish destined for the "usual and accustomed" tribal fishing grounds.

During this period there were many violent confrontations between Indian and non-Indian fishermen as tribal members attempted to exercise their treaty fishing rights. White fishermen, viewing the tribal fishery as unregulated, were bitter about giving up what they saw as their share of the fishery. In 1979, the U.S. Supreme Court upheld the Boldt decision and the healing process slowly began.

In 1983, the federal court ordered the four tribes involved in *U.S. vs. Oregon* and the states of Oregon and Washington to negotiate a comanagement program for the fisheries on the Columbia River. In 1988, the two states, federal agencies and the four tribes agreed to a detailed harvest and hatchery program, taking the parties full circle from confrontation to cooperation.

Unfortunately, at the same time that Indian treaty rights were recognized, the salmon were declining. Runs of more than 8 million fish as late as the early 1950s have dropped to about 2 million today. Within a generation, Indian fishermen watched an abundant resource drop toward extinction.

Legal rights have not protected the tribes from criticism from sport fishermen, particularly in Idaho, where sport fishermen have been unable to fish for salmon since 1977. These fishermen look downstream and see tribal fishermen taking thousands of fish while they are allowed none. When the comanagement program was negotiated, the tribes insisted that Idaho be given no more than an advisory role. Tribal members remain bitter about the state's past indifference toward preserving the salmon resource it shares with its downstream neighbors, demonstrated, they say, by support for the Hells Canyon dams and by intransigence in increasing flows by limiting irrigation. While the tribes blame Idaho's upstream dams for declining salmon runs, Idahoans look downstream and see gill nets, drift nets and downstream hydroelectric dams as the reason for declining runs. It is a conflict that keeps two groups with a similar interest—restoring salmon—apart.

"The state of Idaho wants fish, but they haven't earned it," said Donna Lopez as she wiped fish slime from her hands. Lopez represents the general view of the four tribes that control the Indian fishery on the Columbia. She believes that "Indians deserve the fish, plain and simple," because, as with Idaho irrigators and their prior appropriation doctrine, they were here first.

74   The dispute over tribal fishing remains one of the major obstacles to restoring salmon, especially in the Snake River. Even though tribal fishermen don't target Snake River salmon, they catch them accidentally in their nets while fishing for fall brights, huge salmon that spawn in the Columbia above its confluence with the Snake. Idaho fishing groups suggest that the tribes simply move their fishery above the Snake River so they catch the fall brights but leave the Snake River fall chinook salmon alone. But the tribes are unwilling to give up their "usual and accustomed" places, because it could set a precedent that weakens their legal position. After years of broken promises and broken treaties, the tribes don't give in easily. Today, the tribes argue, they are closely regulated, with seasons set by the states and tribes based on the number of fish that escape through the dams upriver.

The Columbia tribes generally oppose endangered species listing but support restoring salmon. The Supreme Court has ruled in several decisions that the act overrides treaty rights. Salmon recovery under the dictates of the act could spark a new legal battle over the issue. Kathryn Brigham, an Umatilla from Cascade Locks, Oregon, and a former board member of the Columbia River Intertribal Fish Commission, says some tribes believe their treaty overrides the act. "I don't think very many tribes support the Endangered Species Act," says Brigham, who fishes herself. "But we want to take action to rebuild the stocks."

The main losers in the court battles of the 1970s concerning Indian treaties were the non-Indian Columbia River gill-netters. These fishermen have been gill-netting salmon for canneries up and down the Columbia River since the late 1800s. They fish the Columbia River below Bonneville Dam and live in historic small fishing villages such as Chinook, Washington, and Astoria, Oregon. "We used to fish year-round," says Les Clark, a sixty-four-year-old, third-generation Chinook, Washington, gill-netter. "Now we get it in pieces and parts."

The Boldt decision's greatest impact was on nontribal Washington gill-netters along the Columbia because the 50 percent of harvestable fish allowed nontribal fishermen was split among them, ocean fishermen and even Puget Sound fishermen. They were allowed fewer and fewer fish until now it is hardly worth leaving the dock for Clark and other gill-netters. "It's hard to make expenses anymore," Clark said after one of the weekly fall hearings on proposed salmon limits for the next week. Now the Northwest Power Planning Council wants the federal government to deal with the problem by simply buying the fishermen out. Under this plan, the government would buy their licenses, boats and equipment the same way it bought out dairy farmers in the early 1980s. But Clark, who also fishes in Alaska, wants to pass his trade on to his children and grandchildren. He, too, wants to hang on to tradition.

The gill-netters share many of the same woes as the Pacific Ocean trollers in towns such as Westport, Washington. These fishermen catch salmon by trolling in the ocean before the fish enter the river. Now they and their communities have been forced to change like the rest of the fishermen up and down the rivers as they are permitted fewer fish to harvest. The loss of salmon is ending a trade that has been handed down from generation to generation since 1912. Most of the trolling today is done for tourists. Charter boat captain Mark Cedargreen remembers when salmon-trolling boats choked the shoreline off Westport. Now there are as many surfers as salmon fishermen. The towering ten-foot waves that roll into Westport's beaches attract former Californians, such as Ricky Young, who have escaped to the Pacific Northwest. Young, forty-nine, owns a surf shop in Bellevue, Washington, and each August sponsors a surfing contest in Westport that attracts thousands of surfers from throughout the region.

Young says that when he started the contest in 1987, Westport and the entire Grays Harbor region on Washington's coast was foundering. "When I came here the Boldt decision was just ready to roll, timber was down, everything was falling apart," Young says. "I thought I'd try to give the locals a shot in the arm."

The same phenomenon is taking place near Hood River, Oregon. The gusty winds of the Columbia Gorge make the area ideal for sailboards on the artificially wide Columbia River above the Bonneville Dam. Just as surfboards are replacing salmon boats in Westport, sailboards fill the Columbia River off the Hood River in the summer as salmon boats did only a decade ago. And the salmon might have an easier time getting down the river today if they could surf or ride sailboards. In fact, their natural river habitat has become so lethal that most of the young salmon make the trip from Idaho to the sea on barges or in trucks. Then, if the salmon survive the huge, open-water drift nets that are miles long, they have to run through a series of

76

fishing areas to Alaska, back south past British Columbia and down the Washington coast past Westport and other communities where netters, trollers and sport fishermen all have a chance at them. If they make it past the Columbia River gill-netters, they have to go through the eight major dams and a river full of Indian nets before arriving in Idaho. If these adults make it through all of the reservoirs and into the Salmon, Clearwater or Snake rivers, where they spawn, a few Shoshone-Bannock fishermen or poachers may get one last shot before they spawn. Then, two to three years later, when the young of these adults leave the spawning rivers to head to the ocean, they face even poorer odds: in dry years as many as 95 percent may die in the reservoirs, be ground up in turbines, be eaten by squawfish or perhaps die of stress or disease when released from the barges or trucks that the Army Corps of Engineers uses to move the young fish past the dams. This practice prompted Governor Andrus to say that the only place in the Snake and Columbia rivers where salmon are safe is in boats. Yet so many of the salmon die when dumped below the last dam that fishery officials wouldn't even barge the last known living sockeye through the treacherous river system. Like California condors, they were trapped to be raised in captivity.

Mark Cedargreen took over his charter business from his father and remembers the wide-open days of the 1950s when charter boaters and commercial trollers in the Pacific near the Columbia River fished most of the year and caught millions of fish. He acknowledges that overharvesting by all groups has contributed to the decline. But since 1960 a series of government regulatory activities, culminating in 1985 with a treaty with Canada, has brought the ocean harvest under strict regulation. Yet many of the precious fall chinook that still spawn in Idaho are taken annually by fishermen targeting other runs. The recovery plan approved by the Northwest Power Planning Council would allow fishermen to continue to harvest up to 65 percent of the returning adults between ocean and tribal fisheries.

Most of the debate concerning salmon and economics has revolved around the losses to industry, farmers and ratepayers. Yet Cedargreen, Clark and tribal fishermen demonstrate that there has already been a serious financial loss tied to the demise of this once great fishery. The loss of salmon fishing, both commercial and sport, has cost the economy of the Pacific Northwest billions of dollars between 1960 and 1980, says Ed Chaney, who authored a 1982 study on the economic effects of the lost fisheries. But with more accurate estimates today of even higher fish losses, Chaney says his study may have been conservative. He estimated that approximately 44 million adult salmon and steelhead were lost due to all kinds of fishing during the twenty-year period at a cost to the region of $6.5 billion.

Even though salmon income has been declining, the fish continue to bring millions of dollars into the economies of Oregon and Washington. The harvest of ocean trollers in Oregon and Washington generated direct and indirect income of $61.8 million in 1988 and $37.7 million in 1989 according to the Pacific Fisheries Management Council, which oversees ocean salmon fishing. The ocean sport fishery in Oregon and Washington generated $19.2 million in 1988 and $24.4 million in 1989. Sport fishermen at Buoy 10, a marker for a popular fishing area at the mouth of the Columbia, generated $7.4 million in income in 1988 and $5.5 million in 1990. There are no records for sport fishing income generated in the Columbia. Gill-netters in the Columbia, both Indian and non-Indian, generated $46.6 million in 1988 and $17.8 million in 1989.

In 1990, an unexpected and unexplained decline in the number of coho and chinook salmon in Oregon waters worried fishermen and fishery officials. Salmon runs in many rivers were far below what was anticipated, and poor catches were reported in the Buoy 10 fishery. The coho catch off the Oregon coast was about 40 percent behind that of 1989. Many Northwest fishermen blamed Asian drift-netters, who place miles of nets in the ocean far offshore. In 1987, the National Marine Fisheries Service found evidence of salmon poaching by some drift-net fishermen. But so far no Columbia River salmon have been found in illegal catches. Oregon officials dismiss drift-netters as the cause of the decline. They say drift-net operations occur thousands of miles from the traditional rearing grounds of Northwest coho and chinook. Mark Cedargreen agrees. He says that bringing the drift-net fishermen under control will have little effect on salmon numbers. "It would be wishful thinking to think that if we eliminate high-seas drift-net fishing, we will have hordes and hordes of fish returning to the Washington and Oregon coasts," he says.

A salmon might swim 10,000 miles in its amazing journey from Idaho to the Pacific and then back again. Few species depend on as wide an area and so many different ecosystems as Idaho's sockeye and chinook salmon. Destruction of any one ecosystem can mean the demise of these resilient fish. Their survival as a species depends on millions of people across the Pacific Northwest, from the potato farmer of Idaho to the electricity ratepayer of Portland.

All species, including humans, are linked by the water on which they depend. The Columbia and Snake rivers bind the Pacific Northwest together. The various groups that compete for the water have historically been divided between upstream and downstream interests. If the controversy concerning the protection of Snake River salmon has done anything, it has muddied these historical differences.

Today, the communities of the Pacific Northwest are integrally linked, economically and ecologically. The political lines that divide diverse groups are becoming less significant as the region's economy becomes more and more tied to the Pacific Rim and other world markets. The real, yet hard-to-define lines that delineate natural ecosystems, both aquatic and terrestrial, are becoming even more relevant as the activities of humans continue to shrink and destroy them.

For the Pacific Northwest, the 1990s will be the turning point economically and environmentally. As the economy matures from one relying on the development of natural resources to one emphasizing the development of human resources, protection of the qualities that make the region attractive will be even more important. If, in these ten years, Pacific Northwest residents and indeed all Americans fail to protect the rivers that mean so much to the region, then it may be too late. For some of the Snake River salmon, these ten years may be their last.

4. Salmon Sacrifice

ED CHANEY remembers when the fish started dying below John Day Dam.

It was June 1968, and he was gathering information for press releases on the opening of John Day. Chaney was a young, idealistic public information officer for the Oregon Fish Commission, and he was supposed to write about the state-of-the-art fish passage technology installed at the dam. The John Day Dam was the latest in a series of giant dams built on the powerful Columbia River to produce cheap power to electrify the Pacific Northwest. It was the latest political plum brought home by the powerful Washington and Oregon congressional delegations, and a celebration was in order.

Vice-president Hubert Humphrey was to be the honored guest for the dedication ceremony on September 28. Humphrey, one of a generation of New Deal Democrats who embraced the Columbia River's public power system, was running for president. Dedicating a dam showed Pacific Northwest voters what Democrats had delivered to them during the past thirty years: both economic development and power to the people. Humphrey's schedule could not be changed, and U.S. Army Corps of Engineer officials said the dam had to be fully operational in time for the ceremony. Pressure also had come from barge operators and their customers, who ship millions of dollars' worth of goods up and down the river and wanted the dam finished so they could start shipping in the spring. Therefore, even though the hydroelectric turbines would not be pumping out kilowatts until July, the dam had been closed and the river blocked by May. For nearly two months, water poured over the new dam's spillway, driving deep into the pool below. According to Chaney, Corps officials knew that the fish ladders were not working, yet they closed the dam and allowed thousands of migrating salmon to be trapped below the dam.

80

Neither the Corps nor Chaney knew that running the water unimpeded over the spillways increased the level of nitrogen in the pool below the dam. Salmon swimming through this nitrogen-supersaturated water die from gas bubble disease, which produces the same effects in the bloodstream of salmon as the "bends" in deep-sea divers.

May and June are the peak period for the wild salmon run in the Columbia River, and in the spring of 1968 hundreds of thousands of salmon were heading upstream to spawn in tributaries throughout Oregon, Washington and Idaho. With the nonworking fish ladders blocking passage, thousands of huge salmon and steelhead were trapped in the pool at the base of the dam. They were dying, first by the hundreds and later by the thousands. By the time the Corps began diverting water through the hydroelectric turbines on July 17, at least 20,000 salmon had died. Years later officials learned that running water through the turbines alleviates nitrogen supersaturation. But in 1968 the salmon were dying and no one knew why. No one knew the actual number of fish killed either. Gerald Bouck, a Bonneville Power Administration (BPA) fishery biologist said in an April 2, 1991, letter to Merritt Tuttle of the National Marine Fisheries Service (NMFS) that the actual kill was impossible to estimate "because a crew of federal employees allegedly buried dead adult salmon" to hide the carnage. Chaney says he lost his innocence as he watched them bury fish and then try to cover up the whole incident.

Chaney is convinced today that hundreds of thousands of fish died that year, including most of the precious summer chinook stock in the South Fork of the Salmon River near his current home in Eagle, Idaho. He took pictures of the salmon and the cleanup effort below John Day, hoping that such evidence in the hands of fisheries officials could alert the region to this travesty. Maybe his proof could slow the drive to build even more dams at the expense of salmon. But his boss, Robert Schoning, Oregon Fish Commission director, gave him strict orders to keep quiet. This was no time to make powerful political enemies.

Chaney couldn't hold his tongue. He secretly passed the photos of dead and dying fish to the Portland *Oregonian* newspaper, which ran a full-page spread on the story. Chaney had asked for anonymity but was credited for the photos. His career with the Fish Commission was short-lived, but his connection with salmon was forged forever.

"The day they closed John Day Dam without fish ladders, it changed my life," Chaney says. "When the reporter revealed my identity I couldn't go back."

The John Day story unfortunately is not an isolated, unusual event. It is, in fact, typical of the various activities that have brought Snake River salmon to the brink of extinction. Since the arrival of white settlers in the Pacific Northwest in the 1850s, the story of their relationship with the salmon has been one of ambivalence and ignorance. Salmon, one of the most readily identifiable icons of the Pacific Northwest, also played a key economic role in the development of the region, setting it apart from most endangered species, since even today a relatively large business community depends on its survival.

In the last century, salmon have been mined like gold and silver until the mother lode is almost tapped out. And like miners in boom towns throughout the West, the commercial fishing communities have been all but forgotten, like the ghost towns next to empty mine shafts. They don't have a well-connected lobbying machine behind them the way the aluminum and timber industries do. The rich and powerful no longer dabble in salmon canneries in Oregon and Washington. The power base today in the Pacific Northwest lies with the forest-products industry, the aluminum industry and manufacturing giants who depend on cheap electricity. Indeed, the demise of the salmon can be traced to the political power of those who benefited at its expense. Victims of early canneries, commercial fishermen, the timber and mining industries, corporate farmers, the huge cheap-electricity lobby and other river users of today, salmon have suffered for lack of votes and political clout.

After the John Day disaster, Chaney joined the National Wildlife Federation as a lobbyist in 1969 and has since held a series of positions with environmental groups and agencies and worked as a consultant, with most of his work revolving around salmon. Today, as executive director of the coalition Save Our Wild Salmon, Chaney is a cynical, sharp-tongued workaholic. The tall, often tired-looking chain-smoker is still fighting to save the remarkable fish that are a symbol of the wild character of the Pacific Northwest. When he isn't showering congressional offices and newspaper reporters with faxes, Chaney is working with ranchers, environmentalists and bureaucrats to devise new programs for recovering salmon. He is one of a handful of crusaders who have survived in a political and social environment almost as treacherous as the salmon's migration route from the ocean to spawning gravel. These citizen environmentalists have continued to fight a series of retreating skirmishes trying to keep Columbia salmon stocks from going extinct. The next decade will tell if Chaney's battle was only a quixotic campaign doomed to failure or the triumph of an environmental hero of epic proportions.

The immediate future of the salmon depends primarily on whether the dams that stand between the ocean and Idaho spawning grounds are operated for the benefit of salmon as well as for power generation, barging and irrigation. Until Snake River sockeye and chinook salmon were declared endangered and threatened, respectively, under the Endangered Species Act, the fish stood in line behind other dam users for water, protection and funding. It was the continuation of a pattern that began in the 1860s, when the first commercial fishermen began exploiting the Columbia's bountiful salmon resource. Long before the Bonneville Dam was built in 1938, salmon numbers had declined significantly.

To understand the complexity of management needed to restore the salmon, it is essential to understand how this once abundant resource was depleted. Following the history of salmon exploitation in the Pacific Northwest is also following the development of a region that now sustains more than 9 million humans. Just as the forests of the Upper Midwest were mowed down by well-intentioned men seeking to build a nation, the people who decimated the salmon were attempting to settle in a harsh land.

We know little about salmon numbers before white settlers entered the region, except to note that the salmon's abundance made it possible for Indian tribes to build entire cultures around it. Officials today estimate numbers of between 8 million and 16 million salmon in the Columbia River Basin before whites arrived, but no one really knows. It could have been more.

Kirk Beiningen, a fisheries biologist who has worked for a variety of federal and state agencies, estimated fish runs through history in a 1976 paper. Beiningen calculated that, based on an Indian population estimate of approximately 50,000, and a consumption rate of one pound of fish per person per day, as much as 18 million pounds of salmon may have been taken annually before white settlers arrived.

Early explorers noted that there were many more fish in the rivers than the Indians could use.

Between 1866 and 1940, the average catch rose to 29 million pounds a year. Ten times during this seventy-five-year period annual landings exceeded 40 million pounds of fish. The catch appeared to stay high for many years due to increased effort, but eventually it began to decline.

The records of commercial fishing catches document the changes in salmon runs due to fishing pressure. Most commercial fishermen from 1866 to 1889 concentrated their efforts on the prime-quality chinook that ran up the Columbia in late June and July. But as this summer chinook population declined, fishermen shifted their fishing effort to less desirable spring and fall chinook. In the late 1880s and into the 1890s they began fishing for sockeye, steelhead, coho and chum salmon. Fishing was easy, using large seine nets, traps, dip nets, gill nets, drift nets or revolving fishing wheels, which tossed salmon out of the river onto a platform. Beginning in 1912, fishermen went to sea in boats trolling for salmon before they entered the river, hooking even more fish.

Barton Warren Evermann, an ichthyologist with the U.S. Fish Commission, conducted studies of salmon throughout the Columbia River Basin from 1893 through 1896. These studies provide the historic, scientific profile of salmon numbers 100 years ago. Evermann's reports, published in the *Bulletin of the United States Fish Commission,* provide scientists today with baseline data on the life cycle of salmon before the tributaries and main rivers were dammed, before logging and mining had changed the rivers—in short, they are the best indicators scientists have of conditions before humans altered the salmon's ecosystem.

Yet even in the 1890s, Evermann reported dramatic declines in the runs. In fact, it was the decreasing abundance that persuaded the Fish Commission to send Evermann from Washington, D.C., to the Pacific Northwest. The U.S. Fish Commissioner stated in 1894 in one of Evermann's reports that it "is beyond question that the number of salmon now reaching the headwaters of the streams in the Columbia River basin is insignificant in comparison with the number which some years ago spawned in these waters."

In 1896, Evermann and Seth Eugene Meeks, a biology and geology professor at Arkansas State University, were investigating good

84

hatchery sites near Cascade Locks, Oregon, so that the commission could try to restore declining stocks through artificial propagation. They observed that spring chinook numbers already were becoming so scarce that the canneries were not opening to process them. But fall chinook runs were amazingly large. The fish were so abundant they were literally scooped out of the river in fish wheels that were placed in channels that were the main migration routes of the fish. The rest were taken in seines, which would get so full that it took more than human strength to lift them. "It would be impossible to land the net by hand, so swift is the current, and frequently the united strength of four horses is barely able to land it," Evermann and Meeks wrote.

In 1911, nearly 50 million pounds of salmon were canned. Even though the number of fish in the river had diminished, fishermen were getting more efficient at catching them. And by 1919, most commercial fishing and canning was taking place in August, since by then even summer stocks no longer arrived at levels high enough to meet market demand and profitably operate canneries. This "high-grading" or targeted harvesting of the fish during their spawning run peaks was not specifically prohibited by early regulations imposed by state and territorial governments beginning in 1877. "Unfortunately, this practice continued long past the time when severe depletions in nearly all runs had already occurred," Beningen wrote.

Hundreds of miles upstream in Idaho, commercial fishermen ran profitable operations on chinook, sockeye and steelhead when Evermann visited them in 1894. Eleven fishermen between Salmon Falls and Weiser, Idaho, reported a combined catch of 90,000 pounds of salmon and 50,000 pounds of steelhead that year. But the numbers had dropped significantly in the 1880s according to Robert E. Conner, a fisherman from Lower Salmon Falls, who had been fishing since 1882. "For the first four or five years after my coming, salmon were abundant . . . there must have been a thousand in sight at one time," Conner told Evermann. "But there has been a great decrease in the last four or five years."

Evermann correctly recognized that the spawning grounds in Idaho were the key to the abundance of salmon in the Columbia River. Millions of adult fish were still returning to the Salmon, Snake, Payette, Boise and Clearwater rivers in the 1890s, when Idaho remained relatively wild throughout. Evermann witnessed Sawtooth Basin lakes and streams "teeming with redfish." In 1881, some 2600 pounds of sockeye were harvested from Alturus Lake alone for mining camps, which were popping up all over central Idaho. As the new century began, new development activities threatened the salmon runs.

85

In 1910, Sunbeam Dam was built downstream on the Salmon River to provide energy for nearby mines. In 1925, Black Canyon Dam was built on the Payette River by the U.S. Bureau of Reclamation to produce power and provide irrigation storage. These two dams prevented all but a handful of sockeye from returning to their spawning lakes, and the populations in the Snake River crashed. The sockeye, lacking spawning habitat below the dams, largely vanished. Sunbeam Dam also isolated off the headwaters of the Salmon River and the chinook-spawning habitat there, along with the sockeye salmon nursery lakes in the Sawtooth Basin.

Beginning with Swan Falls Dam in 1907, other dams were built that cut off passage to the headwaters of many of Idaho's important salmon streams. Swan Falls, built near King, Idaho, upstream from Boise, stopped the migration of salmon to hundreds of miles of stream east to Shoshone Falls. The Boise Diversion, built by the Bureau of Reclamation to provide farmers with irrigation water for Idaho's growing potato industry, dramatically reduced salmon runs into what was once a key spawning ground for spring and summer chinook. Black Canyon Dam also choked off the chinook runs into the Payette River.

Sunbeam was breached in 1931 when unknown parties ran a dynamite-laden raft into it. Game wardens said there was strong animosity toward the dam by fishermen disturbed not only by its effects on sockeye salmon but on chinook salmon, which also were depleted above the dam. Investigators never found out who blew the dam. Once the barrier had been lifted, an apparent miracle occurred: sockeye salmon began showing up again in Redfish Lake. By 1955, some 4361 adults were counted during spawning season. But the success didn't last. As the four lower Snake dams were built from 1960 to 1975, the Redfish Lake run, now the only sockeye run on the Snake River, dropped steadily to today's single-digit counts. The mangled hulk of Sunbeam Dam remains in place on the Salmon River, a monument to past and present failures of dam builders to protect the needs of the fish.

Mining had started in Idaho in the 1860s, and from that time forward, miners eliminated entire runs into dozens of tributaries

86

throughout the region. Dredge miners, who dug up entire river channels, altered the stream channels or removed the gravel beds where
salmon laid their eggs. Siltation covered hundreds of miles of spawning
gravel with powdery tailings that hindered egg survival. In a June 23,
1911, story on a drowning death, the Grangeville newspaper said the
mines along the river had to be shut down so that the river could clear
enough to find the body.

Mining for copper and cobalt began in the Panther Creek drainage
near Salmon, Idaho, in the 1890s and continued intermittently into the
1980s. In that time countless large fish kills were documented as
tailings pond dams failed, pouring acidic waste rock directly into the
river. No salmon have spawned there since 1967. At first glance today,
as you drive up Panther Creek from the Salmon River, it appears to be
an attractive mountain stream as it bubbles and dances through some
of America's most beautiful landscape. Only when you get to its upper
reaches by the abandoned mining town of Cobalt do you begin to see
the ghastly reds and grays of heavy-metal tailings. The beauty of its
lower reaches is deceptive, though, since the creek has neither fish nor
aquatic insects. Once as many as 2000 adult salmon spawned in
Panther Creek.

After the miners came the farmers, who found Idaho's volcanic soils
ideal for crops such as potatoes, wheat and sugar beets. All they needed
was to transport water to the fields. Irrigators diverted water off many
tributaries, often drying up stream beds just at the time that salmon
were arriving to spawn. Also, millions of salmon would swim up the
extensive network of diversions crisscrossing the Idaho and Oregon
landscapes only to be left stranded in farmers' fields.

The first river Lewis and Clark crossed when they stepped into the
Columbia River drainage in 1805 was the Lemhi River east of Salmon,
Idaho. This remains the eastern boundary of the salmon migration,
nearly 700 miles from the Pacific Ocean. By the early 1950s, a series of
eighty-five separate irrigation ditches had depressed the salmon run on
the Lemhi to only eighty-five fish, and the Lemhi salmon appeared
doomed to extinction. But a program designed to screen the diversions
in the 1950s was a huge success, and by 1961 the run had jumped to
more than 4000 returning adults. Unfortunately, low water and Columbia and Snake dams have depleted the Lemhi run down to near-
extinction levels again.

Cattle and sheep ranchers came with the crop farmers and ranged their cattle throughout salmon country in numbers that overgrazed the forage, causing erosion that filled many streams with sediment. Loggers also cut trees right up to the banks of streams, sending sediment to 87 cover the gravel where salmon once spawned. Often a combination of mining, irrigation, overgrazing and logging led to the ruin of salmon spawning grounds. Bear Valley Creek, a tributary of the Middle Fork of the Salmon River located twenty miles west of Stanley, Idaho, is a typical example. The creek runs out of the central Idaho mountains west of the Sawtooths through ponderosa pine forests. It was once a major spawning ground for spring chinook, which when spawning were so thick that settlers could walk across them. Dredge miners dug up the bed of the creek in the 1940s and 1950s, leaving tailings that continually drained acidic pollution into the stream. The creek was routed away from its original channel during mining, which prevented any spawning for several years in some sections. Thousands of cubic yards of sediment washed down the creek, covering most of the remaining spawning gravel. Then overgrazing removed much of the plant growth that prevented the banks from eroding. This combination prevented what was once one of the most productive spawning areas for spring chinook from recovering.

Meanwhile, the heavy commercial harvest continued. John Cobb of the U.S. Bureau of Fisheries wrote in a report in 1917: "Man is undoubtedly the greatest present menace to the perpetuation of the great salmon fisheries of the Pacific Coast. When the enormous number of fishermen engaged and the immense quantity of gear employed is [sic] considered, one sometimes wonders how many of the fish, in certain streams at least, escape."

Despite decades of overharvesting, salmon catches remained high until about 1930. Scientists don't know exactly why, but from 1930 on, the harvest started to dive. It may have been that the cumulative effects of years of overharvesting and the weakening of stocks from high-grading had finally caught up with the Columbia salmon. Between 1916 and 1920 the average annual catch of chinook salmon alone was about 30 million pounds. From 1941 to 1945 it had dropped to 22 million pounds.

Meanwhile, Franklin Delano Roosevelt and others had a new vision of the value of the Columbia River. Roosevelt, campaigning for the presidency in Portland in 1932, outlined an aggressive plan to harness the Columbia for electric power production. In 1933, he approved construction of the first two major Columbia River dams: Bonneville, completed in 1938, and Grand Coulee, completed in 1941.

88   Roosevelt and the empire builders who designed and constructed the elaborate hydroelectric network in the Pacific Northwest had visions of providing cheap electricity to rural residents. Their dream was largely realized as the power grid spread to even the most remote reaches of the region. The cheap hydroelectric power also helped win World War II by powering the plants that manufactured aluminum for waves of bombers and nuclear reactors that turned out plutonium for atomic bombs. Suddenly the Pacific Northwest, which had been isolated economically from the rest of the nation and exploited primarily for the benefit of the financial centers of the Midwest and the East, finally had its own economic muscle. The power dreamers made taming the wildest section of the United States economically feasible and made the Pacific Northwest more economically independent. Yet the dams they erected were engineered only minimally to preserve the bountiful Columbia salmon runs. These runs supported a multimillion-dollar fishing industry and helped augment the income of rural residents all the way east into central Idaho. Some of the dams, such as the massive Grand Coulee Dam and the Hells Canyon dams on the Snake River, were built without fish ladders, effectively blocking all upstream migration to thousands of miles of spawning streams. Other dams were built without fish passage facilities to allow the downstream migration of salmon smolts in the spring. Even the best-built dams block much of the downstream migration as smolts get lost in reservoirs, chewed up in turbines or eaten by predators after plunging over the spillways.

Eleven of the eighteen dams hindering migration through the Snake and Columbia rivers were built by the Army Corps of Engineers, which operates the dams in conjunction with the Bonneville Power Administration. The BPA is a huge power-marketing agency that reaches into the economy and environment of the Pacific Northwest as few other federal agencies do anywhere else. Created in 1937, the BPA supplies half of the Pacific Northwest's power, serving Oregon, Washington, Idaho, western Montana and portions of Wyoming, Nevada, Utah and California.

The agency sells power wholesale to public and private utilities from a system of thirty dams on the Columbia and Snake river systems, two nuclear power plants and coal-fired plants in Washington and

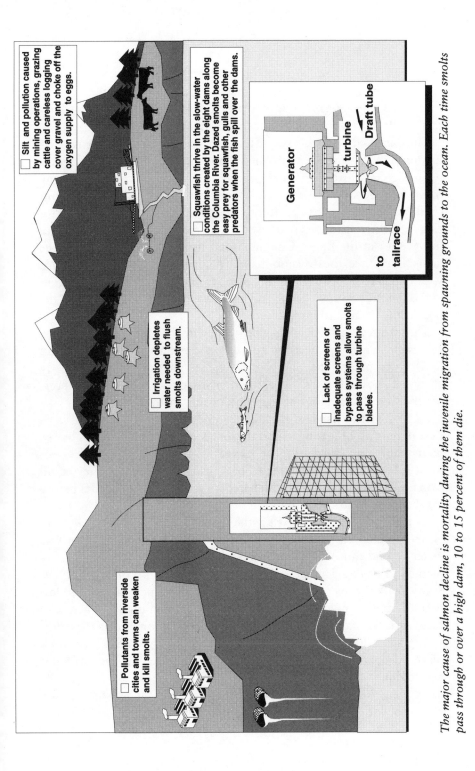

Silt and pollution caused by mining operations, grazing cattle and careless logging cover gravel and choke off the oxygen supply to eggs.

Squawfish thrive in the slow-water conditions created by the eight dams along the Columbia River. Dazed smolts become easy prey for squawfish, gulls and other predators when the fish spill over the dams.

Generator

turbine

Draft tube

to

tailrace

Irrigation depletes water needed to flush smolts downstream.

Lack of screens or inadequate screens and bypass systems allow smolts to pass through turbine blades.

Pollutants from riverside cities and towns can weaken and kill smolts.

The major cause of salmon decline is mortality during the juvenile migration from spawning grounds to the ocean. Each time smolts pass through or over a high dam, 10 to 15 percent of them die.

90

Montana. The system currently generates an average of 8353 mega-watts and has a capacity of 23,665 megawatts, enough to power twenty-four Seattle-size cities. The agency provides 45 percent of the electricity that the Pacific Northwest's more than 9 million residents use.

The dams' lack of fish passage facilities was not entirely an engineering mistake. Bonneville Dam, like Grand Coulee, had no fish ladders planned. Only after fishermen warned of the threat to the economy from the reduction of salmon did the Corps install ladders. The Mitchell Act, which was passed at the same time as the legislation authorizing the BPA, authorized a series of hatcheries to mitigate the damages to fisheries in the upper Columbia and its tributaries caused by the dams. In essence, the decision-makers planned to replace the wild runs with hatchery fish. But even this plan was never fully funded. It went without funding until 1948 and then was funded only because the politicians wanted to build more dams.

This strategy of replacing wild runs with hatchery fish appeared successful when the decision to cut off the runs upriver of Grand Coulee was made. The plan was to relocate the runs, which formerly spawned above Grand Coulee, to other streams downriver by capturing the spawners, collecting eggs and raising them in hatcheries.

By 1948, these fish had successfully added to the Columbia River's overall returns, offsetting in part the losses at Grand Coulee. But like so many short-term successes, the success of the Grand Coulee experiment lulled many of the fisheries biologists into believing they could offset the losses from dams with hatcheries. It would be decades before this bedrock belief was eroded.

In the early years following the construction of Bonneville and Grand Coulee, salmon numbers did decline. Dam construction was given as one cause by biologists at the time, but overharvesting remained in most scientists' minds as the major cause of the declines. Since 1940, the average annual landings of all salmon dropped to less than 12 million pounds. In the late 1940s, salmon numbers began to recover as the harvest was brought under regulation and as measures to improve salmon habitat, such as screening of irrigation ditches, took place. Also, at the time, only one dam stood between the Pacific Ocean and Idaho.

As plans for a flurry of new dams were being made, some voices began to warn against the consequences. The director of the U.S. Fish and Wildlife Service made these risks clear in a 1946 report of the Pacific Northwest Regional Commission.

"It is the oft repeated thesis of the Fish and Wildlife Service that the losses imposed by successive dams are cumulative to salmon migration both upstream and downstream.

"If we are successful in passing the fish over the proposed new dams on the main stem of the Columbia, we will do so with an indeterminate but significant loss. If these survivors are then confronted with a series of four dams on the Snake, there is the strongest doubt that these obstacles can be overcome.

"There is virtual assurance that only a fraction of existing runs could be gotten to the spawning grounds in the Snake River system and that the progeny of this fraction would suffer further loss in its return movement to the sea."

Soon after, in 1953, McNary Dam was completed on the Columbia. The Dalles was finished in 1957, Priest Rapids in 1959, Rocky Reach in 1961, Wanapum in 1963 and finally John Day in 1968. On the Snake River, Ice Harbor was finished in 1961, Lower Monumental in 1968, Little Goose in 1970 and Lower Granite in 1975. With each new dam, the number of returning fish steadily dropped.

Howard Raymond, an NMFS biologist, wrote in 1979 that until 1967, Snake River salmon had only four dams to negotiate during migration. For the most part, survival of juveniles during this period was relatively high despite losses at each dam. But as new dams were added, particularly on the Snake, the trip became harder and longer. The survival of wild chinook salmon from the Salmon River to Ice Harbor Dam, the last one before the Columbia River, averaged 89 percent from 1966 to 1968. But after two more dams were added, average survival dropped to 33 percent between 1970 and 1975. When Lower Granite was finished in 1975, the salmon smolts had to travel through 300 miles of slack water to get from the Snake River to the Pacific Ocean. The trip takes them about a third longer than through a free-flowing river, often as long as sixty-five days from spawning grounds to the ocean. The NMFS said from 1984 to 1990, 86.4 percent of the Snake River spring and summer chinook smolts died before reaching the ocean. The NMFS was hoping to reduce that mortality to 83 percent through measures it approved under the Endangered Species Act in 1992, though these were not enough, it acknowledged, to stop the decline in salmon numbers. Low water conditions in 1992 probably caused mortality to increase. During the drought year of 1977, for example, fisheries officials estimated that more than 95 percent of the smolts died during the downstream migration.

The nitrogen supersaturation problem that killed so many fish at John Day was partially alleviated by placing wings below the spillways to spread the spill through the pool. But even today, there are still losses of adult fish at each dam.

"The hydroelectric system on the Columbia is one of the greatest engineering achievements in the history of mankind," Ed Chaney says. "The dams also are one of the greatest engineering screw-ups in

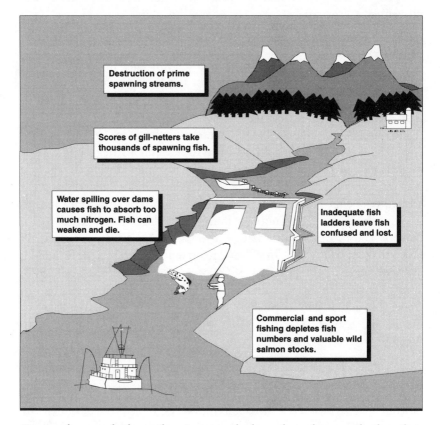

Humans have made the perilous journey of salmon from the ocean back to their spawning streams nearly impossible.

history. There simply was no effort to design the dams to get juvenile salmon downstream."

The Northwest Power Planning Council conducted a study in 1986 to determine how much of the historic salmon losses are attributable to the hydroelectric dams. It determined that 7 million to 8 million fish, or from 50 percent to 80 percent of what it estimated to be the historic run, has been lost permanently because of hydroelectric dams. The council's calculation was done to determine how many fish could reasonably be supported in the basin if salmon were to be fully recovered. It made the assumption that losses due to other factors, such as irrigation, logging, mining, grazing and pollution, are "largely reversible." The runs above Grand Coulee and Hells Canyon dams are gone forever, the council report assumed. And losses at each dam are a given, a situation that never will be fully mitigated.

BPA officials are quick to note that the dams are not the only problem. They and electric utility executives acknowledge the role the dams have played in the decline of salmon, but like each group in the region, they are hesitant to offer bold remedies to end the salmon losses from the dams. "All who contribute to the problem must participate in the solution" has been the rallying cry of the BPA and its supporters. Chaney and other environmentalists react to this strategy cynically, seeing it as an effort to divert attention away from the dams. But others, such as Bill Bakke, executive director of Oregon Trout, grudgingly agree that more has to be done than simply fixing the dam problem. After all, the Columbia River coho didn't have to pass through any dams to reach its spawning grounds; yet the NMFS ruled in 1991 that it had gone extinct.

A recent study shows that logging, grazing and other human activities along Columbia River tributaries have resulted in the loss of 50 percent to 75 percent of the large pools needed by young salmon as rearing habitat. The study, conducted by James Sedell and Fred Everest of the U.S. Forest Service, compared data collected for 300 miles of streams in 1990 with data collected for the same streams between 1936 and 1942. In drainages within wilderness areas or where little human activity took place, there was little or no change in the number of pools, the researchers said.

In the early years of this century, Idaho's trees were, for the most part, left unharvested. Consequently, the salmon streams in the forested areas remained natural where they were not adjacent to mines or

94

grazing areas. In 1932, only 1 million board feet of timber was harvested in the state. But after World War II, the Forest Service, which controlled most of the forested land, actively encouraged timber companies to build lumber mills in the small forest towns of Idaho. By 1959 the harvest had grown to 30 million board feet. In 1983, some 600 million board feet was cut. The harvest peaked in 1987, when 1.7 billion board feet was cut. In the late 1980s, loggers were cutting about 388 million board feet of timber from the forests where salmon streams flow in the Snake River Basin.

Logging can destroy salmon spawning habitat in a variety of ways. Felled trees cast no shadows, so stream temperatures rise and holding areas are lost as the shade is removed. Once the trees are cut, there are no roots to hold the soil. Sedimentation from erosion is the best-known effect, covering spawning gravel and filling in pools. Once the gravel is covered, female salmon are unable to build a nest where the eggs will be safe. Without pools to rest out of the current, the salmon, which are already under considerable stress, will suffer more, and many will die before spawning.

The most dramatic effects of logging on salmon habitat took place on the South Fork of the Salmon River. This huge tributary of the Salmon supported Idaho's largest summer chinook salmon run. Historically, 10,000 summer chinooks spawned in the South Fork. Today, only a handful return.

The granitic soils and steep hills along the South Fork are highly erodible, and the Forest Service was ignorant of this problem when it opened the area to massive logging in the 1940s. More than 700 miles of logging roads were built, 69 percent of them on grades steeper than 45 percent. By 1965, some 15 percent of the timber in the South Fork's watershed of more than 2000 square miles had been removed. Scientists estimated that soil erosion increased by 350 percent. Heavy rains in 1964 and 1965 saturated the soil and the banks tumbled into the river, covering many prime spawning areas with as much as four-and-a-half feet of sediment. In 1965 the Forest Service placed a moratorium on road building and logging within the drainage. Initially, the moratorium allowed the South Fork to push much of the sediment down into the Salmon River. But with the number of returning salmon reduced dramatically by the hydro dams on the Snake and Salmon rivers, the population has never recovered. A major rehabilitation effort begun in 1980 to clear up the remaining sediment has largely been a failure, and the river has reached an equilibrium of sediment loading far higher than before the logging damage of the 1950s and 1960s.

Don Chapman, a fisheries biologist and consultant from Boise, Idaho, who has done exhaustive studies on Columbia River salmon for Pacific Northwest utilities, says the South Fork is a prime example of how logging can ruin a salmon stream. "This is a sad reality that must be occurring in many other salmon-producing watersheds in the Snake River Basin," Chapman wrote in a 1991 report. "Once a stream or river is damaged by logging, the effects may last beyond a lifetime."

The Idaho Department of Fish and Game has sockeye blood on its hands. Its own indifference and mismanagement demonstrate that even those charged with protecting the salmon have played key roles in its decline throughout the century. In the 1950s and 1960s, Fish and Game poisoned sockeye nursery areas in Stanley, Hellroaring, Yellow-belly and Pettit lakes and then installed barriers to keep sockeye from using them again. The lakes remained toxic as long as two years after treatment, then they were stocked with rainbow and cutthroat trout, which the department felt were of more interest to fishermen.

In 1975, Idaho's Fish and Game Commission, which oversees the Department of Fish and Game, voted to try restoring the sockeye run. Working with the NMFS, the department planted hundreds of thousands of Canadian sockeye fingerlings in Stanley and Alturus lakes from 1980 to 1983. The plantings were a complete failure. The NMFS's regional chief of enhancement operations wrote his superiors that the problem was the Idaho Department of Fish and Game. "This program has obviously been given a fairly high public profile in Idaho," he wrote, "but does not seem to be getting the kind of support from the state fisheries agency that is necessary to make sure it has some reasonable chance to succeed."

The Shoshone-Bannock tribes, which petitioned to have sockeye salmon listed as endangered, blame the NMFS as well as the Idaho Department of Fish and Game for not aggressively supporting the fading species. "Both agencies essentially stood by as this species crashed," says Sue Broderick, the fisheries biologist who wrote the tribes' petition.

The rising numbers of hatchery fish in the Columbia River salmon population had blinded the Pacific Northwest's politicians and bureaucrats to the decades-long decline in wild fish stocks. They had spent most of the salmon rehabilitation funds on hatcheries, using them as a narcotic that dulled their sense of reality about the harmful effects of the giant dams they had constructed.

The Shoshone-Bannock tribe finally broke through the cloud of denial on April 2, 1990, when they petitioned the NMFS to list the Snake River's sockeye salmon as endangered under the Endangered Species Act. Two months later, Oregon Trout and three other environmental groups petitioned to list four other salmon under the act— Snake River spring, summer and fall chinook and lower Columbia River coho. In April 1991, sockeye were listed as endangered. In March 1992, the NMFS declared spring and summer chinook to be one species and listed it along with fall chinook as threatened. Coho were ruled extinct because only nondistinctive hatchery stocks were left. The people of the Pacific Northwest finally had to accept the possibility that salmon could become extinct. Now, armed with the truth, they were prepared to begin the painful recovery process. However, like addicts faced with reality, people first reacted with fear.

Suddenly, Pacific Northwest politicians, industries and the media were paying attention to salmon and to the presumed economic disaster their listing could bring down on the region. Only two years after the spotted owl had become a household word, the Endangered Species Act was again at center stage in the Pacific Northwest.

The act could put salmon protection above all other uses of the Columbia and Snake rivers. Flows through the eight federal hydroelectric dams between Idaho and the ocean would have to change. The Bonneville Power Administration, Army Corps of Engineers, Bureau of Reclamation and Pacific Northwest utilities would have to revamp the largest coordinated hydroelectric system in the world and bring the complicated treaties and contracts that govern it into compliance with the act. It was a nightmare for the bureaucrats and politicians who had supported the status quo for so long. But for Bill Bakke, it was finally a chance to get his message heard.

Bakke, a large, soft-spoken, pipe-smoking Oregonian, has been the Pacific Northwest's conscience on the salmon issue for twenty years. For most of that time he was largely ignored by the Portland power structure that has controlled salmon management since the late 1970s. All that changed in 1990 when Bakke's group, Oregon Trout, spearheaded the petitioning for endangered species.

"My sense of Bill, when I first met him, was that he was the Rodney

Dangerfield of the fish business," says Bruce Lovelin, executive director of Northwest Irrigation Utilities, which represents irrigators who also produce electric power. "He was fighting a lonely battle for the wild   97 salmon and no one was listening. I witnessed it firsthand."

Through the late 1970s and early 1980s, Bakke waged his campaign as the voice for wild, naturally spawning salmon. He had stood up to state fisheries biologists, federal fisheries officials and Indian fishermen, who until recently have stridently defended hatchery production. Hatcheries produce fish in the Pacific Northwest specifically to offset the losses to fishermen from dams. So organizations whose constituents are primarily fishermen have fought to keep the hatcheries pumping out salmon for the nets and hooks. At the same time, Bakke says, they are producing "wimp" fish that are slowly watering down the genetic base on which the Pacific salmon population is built.

"Ask yourself a question: Are wild turkeys identical to supermarket turkeys?" Bakke asks rhetorically. "A live supermarket turkey cannot last more than a few days in the wild. The same holds true for hatchery fish. The hatchery product is tame, unadapted, weak, sick and pale. It fares poorly in the natural environment and nature selects against it. It's fine for restaurant fare but not much else."

When you look at the trend you can see why Bakke is worried. In 1970, hatchery production exceeded natural production in the Columbia River Basin for the first time. By 1980, 75 percent of the entire run of salmon and steelhead was hatchery-produced. Today, less than 10 percent of the run consists of wild fish. This worries Bakke as much as the losses in numbers. As the stocks become weaker and inferior in hatcheries, their survival in the wild is bound to drop, he says.

Bakke looks out-of-place in the conference rooms and hearing halls where salmon policy is made in the Pacific Northwest. He looks the part of the fisherman you might expect to find standing knee-deep casting in the Deschutes River or some other salmon stream near his home in Portland. If he looks familiar it's probably because you have seen him before. For much of the 1970s, Bakke modeled for the Portland-based Columbia Sportswear Company. You would find him with his bearded face and big-chested body in outdoor magazines wearing fishing vests and raincoats. During those years he lived a life of leisure. He painted water-colors, nosed through technical files to educate himself about salmon and fished.

98

Over the years, Bakke has taken the state, federal and tribal fish managers to task for allowing salmon to be pushed aside for electricity production, irrigation, logging and commercial fishing up and down the river. Like most of those who have been fighting to preserve salmon, he has counted few major victories. He has brought the Oregon Fish and Wildlife Department around to his side at least partially. In 1978 it instituted a wild fish policy, which requires fishermen to release any wild steelhead caught, and similar policies have spread to Washington and Idaho as well.

Bakke finally began getting the respect he deserves in 1990, when Oregon Trout petitioned for Endangered Species Act protection for chinook and coho. Suddenly, Bakke was one of the most important people in the Pacific Northwest. The petition wasn't filed lightly. Bakke had chosen over the years to work on the edges of the system to steer it with logic and persuasion, but by 1990, he knew the salmon couldn't wait any longer. The region, he claimed, was putting all the risks on the salmon and accepting none for the economy.

"Risk to salmon is acceptable; risk of not meeting electrical load is unacceptable," Bakke says in explaining the mindset he was fighting. "The mute resource, the salmon, bears the risk, and now they are near extinction. No science is certain, but the real question is which way do we err—in favor of the fish or against the fish?"

As Bakke's petition was putting regional economic interests in an uproar, his gentlemanly approach and his avoidance of rhetoric kept him from becoming radicalized in the view of his opponents, even though his positions were a direct challenge to their power.

"He did not pull the endangered species trigger without a lot of deliberation, and I respect him for that," said Lovelin, whose irrigators and power producers face dramatic operational changes as a result of the listing.

Whereas Ed Chaney has concentrated his attack on the hydroelectric dams and the issue of migration, Bakke has focused on the big picture. He has attempted to force the Pacific Northwest to look at the entire salmon ecosystem, from spawning gravel to ocean and back. His ecosystem approach is perhaps the most threatening to the powers that be, for he points out the folly of the technologically based assumptions that people can control the rivers and the hydrologic cycle without destroying the natural system on which it is built.

"The salmon ecosystem developed over millions of years of natural trial and error," Bakke says. "Man's time on the scene is minuscule by comparison. Our trial-and-error tinkering with the ecosystem is leading to disappearance of the fish. Man-designed attempts to mitigate damage to nature are never reliable. Our best engineering minds cannot produce a reliable system as productive as the natural ecosystem."

It's the same story for salmon as it is for most endangered species. If we are to save the salmon we must save its habitat. The most important wild salmon habitat lies in Idaho. It is the Middle Fork of the Salmon River. The salmon that live in this wildest of western rivers are special animals, physically and genetically.

The Middle Fork and its tributaries, such as Marsh Creek, located fifteen miles east of Stanley, Idaho, are the only salmon waters in Idaho that have remained off-limits to weaker hatchery stocks. No hatchery salmon have been mixed in; the evolution that has prepared the wild fish to make the long journey from the Pacific Ocean has ruled unchanged for centuries.

The Middle Fork of the Salmon is among the wildest rivers in America. More than 90 percent of its almost 2-million-acre watershed is within the Frank Church–River of No Return Wilderness. Its 100 miles of pools, riffles and whitewater rapids are boated by nearly 10,000 recreational rafters and fishermen each year. Half of Idaho's outfitting and guiding economy derives from this one river. No one who floats the Middle Fork ever forgets it.

It is the most pristine large river left in the West. Its clarity and purity support not only its economic value, but also its habitat value to salmon. The watershed within the wilderness provides spawning and rearing habitat unmatched in the Columbia River Basin.

The river also has spiritual importance. The Middle Fork was home to ancestors of today's Shoshone-Bannock tribes and is part of their reserved fishing territories. For Idahoans in general, the Middle Fork is a special place. Outfitter Dave Mills spends three months each year on the river. "Every trip is different," he says. "The weather, the water, flora and fauna, the fish—every part of it changes every time. You pay attention every moment. And you connect back—we sleep where those old Indians slept."

Marsh Creek is where the Middle Fork watershed begins. Even though the Frank Church–River of No Return is the largest wilderness

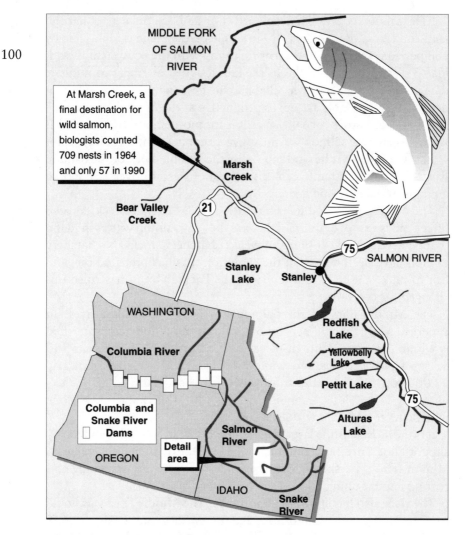

The headwaters of the Middle Fork of the Salmon River and the main Salmon River make up the Stanley Basin, a small area of central Idaho that plays a large role in the future of Snake River salmon.

area in the Lower 48 states, the Middle Fork's headwaters in the Marsh and Bear Valley creeks are one of the few parts of its watershed where roads were built before the wilderness area was created in 1980.

When Terry Holubetz, fresh out of college, first saw Marsh Creek in 1964, the stream was filled with spawning chinook. Today he is in his fifties, with more than a few scars from a career devoted to salmon. Marsh Creek also bears scars of change. Silt, sloughed into the stream by countless hooves of grazing cattle, covers gravel beds that once were 101 prime spawning beds for the red and black chinook. Algae is thick in some areas, a visible sign of poor water quality.

But the big change is the disappearance of most of the large spawners that filled the creek in 1964. That first year Holubetz counted 709 reds. In 1990 he could only find 57—barely better than 1989, when 44 reds were counted. Also gone are the huge schools of juvenile salmon that once filled the pools and riffles of the mountain meadow stream.

"This used to be one of the most heavily used salmon areas in the Snake River drainage," Holubetz says. "It's depressing to see the low numbers of fish and the poor condition of the habitat. I have seen a serious degradation in twenty-six years."

Holubetz has tried to do something about it, first within the Idaho Department of Fish and Game and then as executive secretary of the Columbia River Fisheries Council. He fought for greater flows on the Columbia and Snake rivers. He lobbied for salmon-protection amendments to the Northwest Power Act of 1980. He fought to raise the concern for salmon within the fishery agencies charged to protect them. But despite the power act's promise of "equitable treatment" for salmon and the expenditure of more than $500 million on anadromous fish in the 1980s, the Middle Fork's wild fish and other Snake River salmon stocks are on the brink of extinction.

Holubetz has suffered too. Although he has more experience with anadromous fish management and politics than anyone else in the Idaho Department of Fish and Game, today he plays only a minor role in the state's salmon program. He is one of the casualties of the region's salmon wars.

In 1990, when Senator Mark Hatfield (R-Oregon) organized what became known as the "salmon summit," it was based in part on Bill Bakke's ecosystem model. All the groups that had played a part in pushing the salmon to the brink and who would have to make changes in traditional economic and social programs were brought to the negotiating table along with Bakke and other environmentalists. The summit was going to review the big picture.

"It should not take a listing under the Endangered Species Act to prove the obvious," Senator Hatfield commented at a mid-1990 congressional hearing on the petitions. "Our management strategies are not working."

The salmon summit consisted of a thirty-seat panel representing diverse interests: conservationists, Indian tribes, electric utilities, water-dependent industries, fishermen, federal dam operators, land managers and governors. It met for five months in various Pacific Northwest locations in an effort to reach consensus. The group produced no comprehensive plan to restore the salmon, focusing instead on assisting the fish in 1991. And even this limited agreement for 1991 began falling apart days after the final summit meeting on March 4, 1991.

Thanks to Ed Chaney, salmon migration, and thus the dams that lie between Idaho and the ocean, got the most attention. To get juvenile fish to the sea faster and with less mortality, Idaho governor Cecil Andrus and fish advocates pushed to get regional approval for Chaney's proposal to draw down the four lower Snake reservoirs in the spring. In 1992, when the Northwest Power Planning Council recommended its regional plan, it included a provision to implement drawdown by 1995 if the proposal is found to be biologically sound and economically beneficial. The drawdown is based on the idea that natural migration and travel times from spawning grounds to ocean evolved over thousands of years; therefore, the closer to natural conditions and travel times, the better for fish. By lowering the reservoirs, travel time can be increased without significantly increasing the natural flow of the river. As with most debates, opinion is divided in part between those, like Bakke, who believe that nature knows best and those, like the Army Corps of Engineers, who believe humans can control natural processes and make them work for humans and fish alike.

Part of the driving force behind the drawdown plan is the unappealing alternative, which would be to drain Idaho's irrigation lands dry or build several huge reservoirs upriver at a high cost. And even those proposals might not be enough to flush young salmon to the ocean in the spring before they transform to saltwater animals.

Since 1985, more than 97 percent of juvenile salmon that have left Idaho for the ocean have been collected at several dams and barged or trucked downriver and dumped in the estuary below Bonneville Dam. Most are captured at the first dam, Lower Granite, or at the second dam, Little Goose. Those salmon that are missed are caught at McNary Dam on the Columbia.

The salmon transportation program is a tacit recognition that their natural river habitat has become so lethal that another method of getting them to the ocean must be tried. Once again the debate focuses not so much on economics but on who does the job better, nature or humans. Environmentalists have never been comfortable about barging or trucking the fish. Studies in 1978 and 1986 touted by the BPA and the Army Corps of Engineers showed the survival of barged fish to be as much as two times greater than that of those flushed through the reservoirs. Steelhead and fall chinook, which are less sensitive to travel time, responded the best. But newer studies, released in the early 1990s, brought into question the validity of the original studies. The fish may have been surviving the trip to the estuary, but they were not returning to the spawning streams in the numbers scientists expected. The more juvenile salmon that were barged, the fewer that returned as adults two years later.

"Since we have been barging nearly 100 percent of all Idaho juvenile salmon for seven years, you would think that if it worked, we would not be putting spring and summer chinook on the Endangered Species List. We would be fishing for them instead," says Pat Ford, a Boise, Idaho, environmentalist and writer.

The Fish and Wildlife Service, which is in the unenviable position of being responsible for inland salmon habitat management but not for overall salmon management, took a middle-ground approach to the issue in a report it made to the salmon summit in 1991: "Available data for spring-summer chinook and sockeye from studies conducted in the 1980s indicate there is no benefit or a negative impact to returns to spawning grounds and hatcheries. The fact that spring and summer chinook in the Snake River are exhibiting signs of stock collapse even with a massive transportation effort supports this conclusion."

Management of the rivers for power and fish was largely divided until 1980, when Congress approved the Northwest Power Act. The act was first promulgated by the region's utilities to better coordinate management of the electrical system between public and private utilities. Soon it became the vehicle for expanding energy conservation and salmon preservation efforts. The act required the BPA, Corps, Bureau of Reclamation and the Federal Energy Regulatory Commission, the agencies that controlled hydroelectric dams in the region, to "protect, mitigate and enhance Columbia Basin fish and wildlife, to the extent affected by the development and operation" of the Columbia's

104

hydroelectric system. It required the federal agencies to take into account fish and wildlife "to the maximum extent practicable" in their operations. The most important line in the bill as far as salmon were concerned was the requirement that the four agencies provide "equitable treatment" of fish and wildlife in all activities not specifically constrained by law. "It is clearly intended that no longer will fish and wildlife be given secondary status by the Bonneville Power Administration or other federal agencies," said Representative John Dingell (D-Michigan), one of the key sponsors of the legislation. This law was seen by fish advocates as a breakthrough. To them it was the hook on which they could hang salmon recovery. But it didn't turn out that way.

In 1982, the Columbia Basin Fish and Wildlife Program was approved for the first time by the Northwest Power Planning Council created by the act. It looked so promising at the time, says Ed Chaney. It recognized for the first time the need to augment flows in the rivers to assist downstream fish passage in the spring. It set a goal for optimum and minimum flows necessary to flush juvenile salmon down the Snake and Columbia rivers. This "water budget" was to come out of water held in the system for hydroelectric generation. The fish bypass systems, which are now used to collect juveniles for barging, also were approved. Finally, a fish propagation plan was proposed that favored wild fish over hatchery stocks.

Unfortunately, the council was not willing to make the tough decisions that less than a decade later it acknowledged were necessary. It rejected a recommendation by state and tribal fishery agencies that flows in the spring be set on a sliding scale, with 85,000 cubic feet per second considered the minimum flow on the lower Snake and 140,000 cubic feet per second considered optimum. That way, fish would have shared the abundance of water along with the shortage, yet still get what was needed. Instead, the council set the water budget at a constant flow of 85,000 cubic feet per second, the minimum. Even then, in the years after it was passed, the water budget minimum was never met through the entire migration period.

The Pacific Northwest's power supply system had faced the 1980s at a time when utility planners were warning of the threat of a power shortage. But by 1982 it became clear that, in fact, the system was facing a huge surplus. That miscalculation had a far greater effect on power rates than any salmon recovery plan ever will have. Paying for the decisions that resulted from this one massive mistake has driven the agencies to ignore the "equitable treatment" doctrine of the Northwest Power Act.

In 1976, then BPA administrator Donald Hodel sent a letter to all of the preferential customers of the BPA, mostly public utilities throughout the region. In the letter, Hodel warned the utilities that based on forecasts conducted by the BPA, by 1983, it could not assure those utilities a firm supply of electricity. Hodel's forecast was based on what was almost a religious belief in a 7 percent electrical power demand growth in the 1970s. That was the level of growth through much of the 1960s and up to the early 1970s.

But increased energy conservation, sparked by the Arab oil embargo of 1973, had dramatically lowered the demand growth to less than 3 percent by the end of the decade. Meanwhile, the Washington Public Power Supply System (WPPSS), which provided public power to much of the urban area surrounding Seattle, had already begun building three nuclear power plants to meet the growing demand. Hodel's letter supported the need to build even more, and two additional nuclear plants were planned in partnership with eighty-eight public utilities throughout the Pacific Northwest. The utilities, spurred on by the BPA, placed themselves in obligation to pay the debt service on bonds to build the nuclear plants. The combination of lower demand and construction delays forced the WPPSS to suspend construction of the third and fourth nuclear plants and halt engineering on the fifth in 1982. Soon after, the bonds on the project went into default. After lengthy litigation, a debt repayment plan was arranged, with a portion of the debt load placed on the BPA. Today, only one of the five plants is operating and a second is mothballed, with little likelihood that it will ever open, even though it is nearly complete.

"The BPA gambled and lost on the WPPSS nuclear plants," Chaney says. "Now they are trying to make the fish pay for their mistakes."

Today, more than $600 million of the BPA's annual budget goes to servicing its debt for the five nuclear plants, or about 33 percent of its total budget. That compares with an annual fish and wildlife budget of $53 million, or about 3 percent of its budget. So even the most expensive plan for saving salmon will represent a small part of the region's energy bill.

So why is the BPA worried about losing a few kilowatts of power potential to fish? First, the surplus is gone. The Northwest Power Planning Council predicts that the region will need from 12 percent to more than double the amount of electricity it has today to meet demand in the next twenty years. That wide gap ranges from low-growth to high-growth scenarios. Any new power is more expensive than hydropower—from about one-third of a cent per kilowatt-hour for hydroelectric-generated power to about five cents for coal or nuclear power.

Also, after the WPPSS debacle, Bonneville figured out a way to pay off its debt while at the same time dealing with the surplus power it had in the 1980s: it sold the power to California, often for less than the price it sells to Pacific Northwest customers. Dan Chasan, author of *The Fall of the House of WPPSS*, said the need to pay off the debt has driven the BPA to be stingy with the water behind its dams. "The WPPSS debt is one of the things driving Bonneville to get every cent it can out of the river," Chasan said in an interview with the *Idaho Falls Post Register* published December 24, 1990.

When BPA deputy administrator Jack Robertson was asked at the 1991 salmon summit about what would happen if major changes were to be made in the hydro dam operations to save salmon, he didn't talk about rate increases. Instead, he warned that such changes would create the need for additional power sources, such as more nuclear power plants or perhaps coal-fired plants. Each of these, he noted, has environmental consequences, such as carbon dioxide emissions from coal plants that contribute to the greenhouse effect. "There is a cost in everything we do," said Robertson.

Chaney and others in the region believe that a well-managed power system and a salmon-safe river system could cost the same as the current system. Jim Lazar, consulting economist for the Northwest Conservation Act Coalition, says the BPA isn't being creative enough to figure out how to save fish and not raise rates. "Bonneville's taking a shortsighted look at the problem. They're not looking at alternative solutions," says Lazar.

During the spring, for instance, the hydro system produces a surplus of energy that is sold to California at low rates. Jim Baker, Northwest representative of Friends of the Earth in Walla Walla, Washington, says managing the Columbia for a drawdown in the spring could make the water available in the summer and winter, when it is more valuable and marketable. Energy exchanges, in which California or Southwest utilities trade power during the winter for power produced in the summer, also could help. The BPA has already begun some exchanges with southern California utilities. The region also has water available from British Columbia that could be used to aid salmon and produce power if managed properly, according to Chaney. The region also could market its power east instead of simply to the north and south, gaining seasonal markets not now available. As coal resources in the West are developed in Montana and Wyoming, seasonal trade-offs with the hydropower system could be advantageous for both areas. When the Portland-based utility Pacific Power and Light merged with Utah Power and Light to form Pacificorp, it was able to take advantage of the mix of coal and hydropower. Pacific Power was a winter-peaking utility—that is, it used most of its power in winter—while Utah was a summer-peaking utility. The BPA can better market power when it needs it the least by opening new markets. Up to now, however, the BPA has resisted any dramatically different marketing options, but the need to solve the salmon migration problem could eventually move the agency to seek imaginative solutions that not only help salmon but ratepayers as well.

Moreover, energy conservation is already playing a major role in preventing the need for expensive new power plants. Even more conservation opportunity is expected to be available in the region rather inexpensively in the next twenty years. The planning council determined that 3200 megawatts of conservation from residential, commercial, industrial and agricultural sectors may be technically available in the next twenty years. Another 700 megawatts of conservation was deemed promising. If all of this fails to meet the overall power need, the council could reverse its recommendation that natural gas not be considered an efficient alternative to electricity for heating houses and industry. This could free up thousands of kilowatts of energy in the short term, since cheap natural gas is available in abundant supply from Canada.

Even if higher power rates are required to save fish, the Endangered Species Act will force the Pacific Northwest to give salmon at least equal consideration with economics. While some short-term fixes are needed immediately, Ed Chaney says that, over the long term, the region should begin plans to rebuild its dams with full fish passage facilities, the way they should have been built in the first place. It is the same lesson that emerges in nearly every story of endangered species in the Pacific Northwest and worldwide. The sooner the region acts to preserve salmon, the less costly recovery will be, both economically and socially. Unfortunately, in the 1990s, the cheapest alternatives are no longer available.

5. Lost Opportunity

THE Reagan administration left both physical and emotional scars on the Pacific Northwest in the 1980s.

The results of the decisions by Reagan appointees and congressional leaders from the Pacific Northwest are shown in the clearcut-pocketed landscape seen from the air as you fly across the states of Oregon and Washington. They can be seen in the faces of the unemployed mill workers, loaded with bitterness from the loss of jobs and a future. The lack of foresight by career resource managers during the last twenty years can be seen in the silt-laden rivers and streams that run into the Columbia River and now are nearly devoid of spawning salmon. It also is displayed along the Oregon and Washington coasts in the fishing villages where boats are sitting in port for lack of fish.

The Reagan administration and Pacific Northwest resource policymakers decided early in the decade that cutting trees, producing cheap electricity, barging grain and many other economic activities in the region not only were more important that preserving fish and wildlife, they were so important that some fish and wildlife species could be allowed to go extinct. Sometimes these were not conscious decisions; other times actions that were taken were blatantly illegal.

U.S. District Judge William Dwyer of Seattle, who was one of the federal judges who finally forced the U.S. Fish and Wildlife Service (FWS) and the U.S. Forest Service to protect the northern spotted owl in 1990, was harsh when, a year later, he had to order the FWS once again to follow the dictates of the federal Endangered Species Act to protect the owl. "More is involved here than a simple failure by an agency to comply with its government statute," said Dwyer, a Reagan

109

appointee himself. The Forest Service and the FWS had combined in "a remarkable series of violations of the environmental laws."

110   He didn't stop there. Dwyer placed the blame at the top levels of the Reagan and Bush administrations. "This is not the doing of the scientists, foresters, rangers and others at the working levels of these agencies. It reflects decisions made by higher authorities in the executive branch," he wrote.

Dwyer was writing only about activities surrounding the northern spotted owl, but he could have been talking about salmon, snails or many other threatened and endangered species in the region that stood in the way of economic development. The Reagan administration was elected to office to loosen the chains that were holding back businesses. In public resource policy that meant opening public lands to development and getting government off the backs of the people and corporations who could turn those resources into jobs and commerce. The result ten years later was not what the timber executives nor electricity czars who had lobbied for such action expected. Their lives and the lives of most Pacific Northwest residents will be more complicated because of the resource decisions of the early 1980s.

In 1981 the FWS decided after an informal review not to list the northern spotted owl as an endangered species. The same year, the National Marine Fisheries Service (NMFS) chose not to list Snake River chinook salmon as threatened and endangered. A decade later both owls and salmon were protected under the Endangered Species Act, and the Pacific Northwest was locked in political and even social turmoil. The FWS estimates that about 30,000 timber-related jobs will disappear because of the listing of the spotted owl. Thousands of fishermen already have put away their nets and trolling poles as salmon numbers dropped. Now the futures of irrigation farmers, aluminum plant workers, miners and others tied to the region's cheap power are in doubt.

What makes this story particularly sad is that it didn't have to happen this way. The Pacific Northwest, especially its rural areas, was destined to go through a transition, but a transition that could have been accomplished with a minimum of pain. As the timber frontier ended at the Pacific Ocean and as the amount of old-growth timber declined, some reduction in the work force was inevitable. But the reductions could have been stretched over a longer period of time, perhaps even into the next century, when larger stands of second-growth timber would mature enough for the creation of jobs and commerce.

What managers and policy-makers failed to do a decade ago was to listen to the warnings from scientists concerning the spotted owl and the Snake River salmon in 1981. The region's forest and river ecosystems were in trouble, and scientists had been reporting as much as far back as the early 1970s. Had reasonable management programs been put in place early in the 1980s, perhaps as a result of following the mandate of the Endangered Species Act, the Pacific Northwest would be better off economically and environmentally today. There would still be timber left to cut and more options for cutting it. There would be more salmon in the rivers and more flexibility for saving them while preserving the economies that depend on the Columbia River. The lesson for the future is that prevention is easier than a cure. Preserving the ecosystems that support the biodiversity of the Pacific Northwest, the United States and the world is easier than trying to restore them later.

When a species is considered for protection, the Endangered Species Act demands that only scientific evidence be considered. In the 1980s, Reagan administration officials ignored this, taking into account everything from politics to economics. Today, in nearly every case, the species are even worse off than they were when first reviewed, and even tougher, costlier measures are needed to recover them, if in fact they ever can be. The Reagan administration's management of endangered species in the 1980s assured that the northern spotted owl and Snake River salmon would become household words in the 1990s.

Rolf Wallenstrom found himself in the middle of endangered species politics in the Reagan years. Wallenstrom, now in his late fifties, joined the FWS in 1959, shortly after graduating from Syracuse University, in his home town of Syracuse, New York. He started his career negotiating with landowners to buy wetlands for the FWS in Minnesota and South Dakota. He directed the FWS's important wetlands program in South Dakota until taking the area manager job in 1970, overseeing FWS offices in South Dakota and Nebraska.

It was there that he ran into his first political controversy. Nebraska farmers and Colorado cities both had dreams of building dams on the Platte River to store water for farming and urban development. Wallenstrom was charged to protect sandhill cranes and wanted to purchase sandhill crane habitat near Grand Island, Nebraska. In the end, the dams weren't built, the crane habitat was protected and Wallenstrom was promoted. "Standing up for what you believed in professionally was not a sin in the Nixon administration," Wallenstrom says.

The agency was headed by a professional biologist who had worked his way up through the system. That was the way it had always been, and the practice had preserved the professionalism that lasted until Reagan took office. In 1981, for the first time, a director was chosen from outside the service. Robert Jansen, a state wildlife department director from Arizona, was given the post as the first political appointee.

At the same time, James Watt, an outspoken conservative ideologue, was appointed Interior secretary. Fish and Wildlife Service directors had always had to deal with politics, but until James Watt came in, directors had been able to shield the rest of the service from the political whims of the different administrations. That made responding to the mandate of the Endangered Species Act—to consider only science in listing decisions—much easier. Watt had other ideas. He came into the Reagan administration from the Mountain States Legal Foundation, a conservative, Colorado-based legal firm dedicated to protecting private property rights and keeping western ranchers in control of land-use policies of the federal government. It was the time of the Sagebrush Rebellion, when some westerners were in open revolt against the implementation on public lands of environmental laws passed in the 1970s.

Watt, a former Department of the Interior official, knew the ropes and knew he had only a limited amount of time to put in place permanent policy changes he hoped would last beyond his tenure. Most of his attention was on land and mineral management agencies inside Interior. But he didn't want the Fish and Wildlife Service getting in his way. He believed it was proper to consider economics in listing decisions, if for no other reason than that the FWS already was not adequately funding recovery of the species it already listed. "I told them that we should be taking care of the species we already have on the list," says Watt. "I wanted them to get more recovery plans done instead of adding more species we were unprepared to manage."

Wallenstrom, who at that time was the FWS's associate director for environment, remembers Watt's directives differently. "I remember the goal was resisting listing and minimizing budget expenditures for recovery also," he says.

In the first years of the Reagan administration the listing program came nearly to a halt. Only four species were listed in 1981, ten in 1982 and twenty-three in 1983. For comparison, in the later Reagan years after Watt left, between 1984 and 1988, the FWS was listing species at a pace of about fifty to sixty a year. The lower listings were due to pressure from Interior. "It wasn't uncommon for the department to challenge and delay," says Wallenstrom. "It was quite obvious they were reacting to political pressure." The delays took their toll. During the 1980s, thirty-four animal and plant species became extinct while awaiting final listing decisions, according to a Department of the Interior inspector general's report released in 1990.

In 1985, the FWS got a new director: Frank Dunkle, a former Montana Fish, Wildlife and Parks director, replaced Jansen. About the same time, Wallenstrom moved to Portland, Oregon, where he took over the job of regional director. Dunkle, trained in a state where political interference with fish and wildlife policy is commonplace, fit in well with the Reagan administration's Department of the Interior. Meanwhile, Wallenstrom was placing himself on the wrong side of administration policy in several areas, including opposing offshore oil drilling and working to reduce serious selenium pollution at the Kesterson Wildlife Refuge in California.

The northern spotted owl case highlighted Wallenstrom's differences with the administration. It was one in which he first went along with a decision not to list the owl, which he knew was wrong, and later refused to cover up the political manipulation of the listing process. The most telling example of the Reagan administration's mishandling of the Endangered Species Act started when Greenworld, an environmental group from Cambridge, Massachusetts, filed a petition to list the northern spotted owl as endangered in January 1987. It was the second time in the decade that the owl was up for listing, but by then even more of the old-growth forest on which the owl depends was gone. In 1987, owl habitat in the coastal forests of Oregon, Washington and northern California was being logged at a rate of about 74,000 acres per year. An accelerating logging program that had merely been a threat in 1981 was now a reality.

Since an outside group had petitioned for listing, a very strict legal process was set in motion. Once the FWS receives a petition it has ninety days to decide whether it has enough data to make a decision, and to officially "accept" the petition. If it decides it has enough data, then it has one year from receipt of the petition to decide whether to propose listing, an action that triggers a public comment period. After the FWS proposes listing a species as threatened or endangered, it has one year to reach a decision whether to list or not. Normally, a decision on whether the information available warranted further study would have been Wallenstrom's and his staff's, but Fish and Wildlife's staff in Washington, D.C., routinely reviewed and still reviews such decisions. Even though Wallenstrom's regional office recommended accepting the petition in April, near the deadline for a decision to accept it, Dunkle held it up in Washington until July, already breaking the law by missing the deadline. Wallenstrom says the delay was not totally the fault of Dunkle. Politics had taken the decision out of his hands. Another Interior official, James Cason, who under normal procedures had no role in the decision, was doing everything he could to manipulate the process.

Cason, who was assistant secretary in charge of the Bureau of Land Management (BLM), was one of the young breed of conservatives brought into the administration by Watt to restructure Interior from the top down, into the ranks. The BLM oversees hundreds of thousands of acres of prime old-growth timber in Oregon. Cason didn't want the northern spotted owl or the Fish and Wildlife Service getting in the way of the BLM's plans to cut as much of the forest as fast as it could. He later gained brief fame when in 1989 his appointment by President Bush as the assistant secretary overseeing the Forest Service was blocked by a concerted lobbying campaign by environmental groups. During a 1987 conversation with Wallenstrom and a Forest Service official, Cason made his view clear. "I'll bet you anything right now you would recommend a listing and you'll do it over my dead body," Wallenstrom quotes Cason as saying. "He was adamant," Wallenstrom adds.

But Wallenstrom wasn't the only FWS official to whom Cason was talking about the owl. Dunkle said several of his staff were getting calls from Cason. Shocked and appalled, Dunkle said he took Cason's interference to the office of Interior Secretary Donald Hodel, a Watt assistant who was appointed to the top Interior job in 1985. "I said listing was solely the job of the Fish and Wildlife Service, and he agreed," says Dunkle. "I was protecting my people in the field."

While Dunkle says he defended Wallenstrom and others from political interference, his own actions suggested he was not immune. The amount of time left for actual study of the petition was reduced when Dunkle set December 1, 1987, as the deadline for reaching a decision to propose listing, which was nearly two months sooner than man-   115 dated by the act. Dunkle says he set the earlier date so that there would be no conflict with the holiday season. When earlier delays and the time used to form a study team were subtracted, the panel had only three months to collect the information needed to make the decision.

Jay Gore, the Fish and Wildlife Service biologist from Boise who headed the study team, says Dunkle had not allowed enough time for a complete review. And that wasn't the worst of it. Gore's job was complicated by an unusual requirement that had come down from Dunkle's office. "We were not supposed to put any conclusions or recommendations in the report," he says.

Gore's team stated in the report that Forest Service harvest plans in the ancient forests, on which the northern spotted owl depends, would lead to its extinction. The report documented the loss of 62,000 acres of spotted owl habitat per year—less than environmentalists said was actually being lost—and found that the reductions in observations of owls were associated with the rate of decline. To strengthen the scientific validity of the report, Gore had outside spotted owl experts review and endorse it before he submitted it to Washington. Gore's report, which was not the message his bosses wanted to hear, added insult to injury by opening with a cover letter that specifically recommended a "threatened" listing for northern spotted owls on the Olympic Peninsula in Washington, where the worst destruction of owl habitat had taken place.

Wallenstrom said FWS officials in Washington requested regional assistance in rewriting the report to support a decision that the owl was not endangered. The section of the report that showed that the proposed harvest plans would lead to the owl's extinction was chopped out and replaced with a "sanitized" version. The peer-review comments also were removed.

All of this was done without informing Gore, who was the principal author of the report. When he finally saw the report, he says he was shocked. Not only did he hardly recognize the report after editing, but it did not accurately reflect the biologists' review or their emphasis on the dangers to the spotted owl population. In short, it was a whitewash, and a sloppy one at that.

Wallenstrom says he expected Dunkle to sign the final decision after

116

altering the report. Instead, two days before the deadline, Dunkle handed the decision back to Wallenstrom. Pressured by his boss, Wallenstrom told Dunkle he would consider listing only the more imperiled owl population on Washington's Olympic Peninsula, or deny listing.

"He asked me what I was going to do, and I said perhaps I would list the part of the population on the Olympic Peninsula or not list at all," Wallenstrom says. "He implied he wouldn't go along with listing on the Olympic Peninsula."

Despite his better judgment, Wallenstrom succumbed to the pressure. He turned down the owl's listing. Dunkle denies intimidating Wallenstrom. He says he supported listing the Olympic Peninsula population from the beginning and would have issued the decision recommending the partial listing if he had made the decision. "It was his decision," Dunkle says of Wallenstrom. "It's his name on the memo."

Others in the FWS challenge Dunkle's assertion that he supported listing. "We all knew that Mr. Dunkle didn't like the Endangered Species Act," said Bill Reffalt, who directed the wildlife refuge program under Dunkle.

Conservation groups immediately challenged the decision in U.S. District Court in Seattle, and Judge Thomas Zilly ruled in 1988 that the decision had not been based on biology. But it wasn't the manipulation of the process that led to the judge's decision, says Melanie Rowland, an attorney with The Wilderness Society who participated in the case. "They simply had no scientific basis for their conclusion," she says. "Usually that kind of a case is difficult because if they had any expert opinion the judge would rule for the agency. But they couldn't find anyone to support them."

The General Accounting Office (GAO), a congressional investigative agency, launched a probe into the spotted owl listing review after Zilly's decision. Its investigators talked to Wallenstrom, who told them the whole story. Soon after, in 1989, Dunkle fired Wallenstrom and replaced him with a political appointee, the first time politics had reached down to the regional director's level. But the GAO's report, released in February 1989, supported Wallenstrom's claims and helped to restore him to his position. Wallenstrom retired later that year.

"Decisions on listing petitions like this one for the northern spotted owl can often be surrounded by highly emotional debates centered on the decision's possible economic consequences rather than its biological merits," wrote the GAO team. "In such cases especially, FWS needs to be able to demonstrate that its review process and ultimate decisions   117 have been as thorough, independent, and objective as possible. There is evidence that the spotted owl process did not meet such standards."

What had the Reagan administration gained from the delays? Plenty for the huge forest-products industry. Between 1987 and 1989 a record 20 billion board feet of timber was cut from federal lands in Oregon and Washington. About 150,000 acres of owl habitat was logged during that time.

In June 1989, the FWS gave up its court battle and proposed listing the owl, and a year later the agency listed it as a threatened species. But the damage had already been done. A decade of haphazard harvesting, at a rate faster than the trees could grow, had left little owl habitat and no long-term future security for many timber workers and the towns they supported.

In 1972, Eric Forsman, a graduate student at Oregon State University, and his professor, Howard Wight, had first warned that the northern spotted owl might be threatened by extinction. When the Endangered Species Act was passed in 1973, the bird was among the first candidates for listing, and an Oregon endangered species task force, which included biologists from several state and federal agencies, including the BLM and the Forest Service, recommended preserving 300 acres of old growth around each spotted owl nesting site. Their idea was rejected by the BLM and Forest Service hierarchy. Today, the two federal agencies have been forced to preserve more than ten times that amount of land around owl nests in some areas.

In 1977, the task force developed the first spotted owl management plan, calling for at least 1200 acres to be protected for each pair of owls, with 300 acres of old growth allowed at the core. The Forest Service grudgingly accepted 1000-acre owl protection zones, but the BLM refused to commit to a specific acreage figure. In 1981, the same year that the FWS rejected listing the owl as threatened or endangered, the Oregon-Washington Interagency Wildlife Committee's spotted owl subcommittee recommended leaving at least 1000 acres of old growth within a 1.5-mile radius of each owl nest, the minimum Forsman said he had seen owl pairs use around a nest.

Forsman told Seattle *Post-Intelligencer* reporter Rob Taylor in 1992 that the committee's recommendation was a mistake. "It just doesn't make sense to manage a species for the minimum amount of habitat," Forsman was quoted by Taylor in the May 1992 issue of *Government Executive* magazine. "That leaves no room for error. Constantly trying to compromise biology with politics doesn't work very well."

The Oregon-Washington subcommittee biologists told the Forest Service that owl zones needed to be within three to six miles of one another, measured from the centers, to allow young owls to recolonize areas left vacant by dead owls. The Forest Service instead approved six- to twelve-mile spacing measured from the edges. The BLM's decision was even worse; it spaced some areas as much as fifteen miles apart.

Bill Ruediger, a wildlife biologist with the Forest Service, was working on owl studies in the Gifford Pinchot National Forest in Washington from 1979 to 1986. He said the pattern of harvest, which did not follow the recommendation of many of the owl biologists, made what old growth that was left in many areas unusable by the owls. With the choice of the larger spacing options, owl zones that were protected were often isolated from other owl zones, so today many of the areas have become little islands surrounded by clearcuts, which owls won't cross.

"There were less severe options available ten years ago that are not available now," Ruediger told the *Missoulian*, a Missoula, Montana, newspaper, in 1990. "The old-growth forests got cut up a little more every year."

For Forest Service officials the pressure was immense. In 1976, Congress passed the National Forest Management Act, designed to balance the need to protect wildlife, fish, water quality, recreation and other values in the national forests with the need for timber. Senator Mark Hatfield (R-Oregon) inserted a special provision specific to the Pacific Northwest that directly challenged the main goal of the act. The provision allowed the Forest Service to exceed the harvest levels its foresters said the forests could sustain. In other words, the Forest Service was allowed to cut more timber than it could grow. This would permit the Pacific Northwest timber industry to cut more timber on federal lands to make up for declining supplies on private land.

The recession of the late 1970s and early 1980s slowed the harvest— so much, in fact, that many in the timber industry petitioned Congress to allow them to back out of timber purchases they had made earlier with only a minor penalty. Later, however, the same companies in many cases would get to buy back many of these sales at a fraction of the cost they originally would have paid, since by then the Reagan administration was joining Hatfield and the powerful Northwest congressional delegation in pushing for higher timber harvests to help the region's economy. At the vanguard of this effort was John B. Crowell, a former vice-president of Louisiana-Pacific Corporation, one of the largest buyers of federal timber. In 1981, Crowell was named assistant secretary of agriculture for natural resources and environment, which put him in charge of the Forest Service.

Crowell was convinced that the Forest Service could double the amount of timber it was cutting annually in the Pacific Northwest— from 5 billion to 10 billion board feet annually—even though this would have taken the harvest far above the level that could be sustained into the future and would have violated dozens of environmental laws.

Crowell didn't get his way exactly, but he did get the Forest Service to increase the harvest steadily through the end of the 1980s. His strategy in the Pacific Northwest was to delay the completion of forest management plans required under the 1976 National Forest Management Act, plans that inevitably called for reducing the harvest. Forest Service supervisors, aware that implementation of these plans would dramatically reduce harvests, pleaded with top Forest Service officials to begin lowering the harvest gradually so that the timber industry and timber communities could ease into the expected period when supplies would be lower before rising with the maturation of second-growth forests. Max Peterson, a former road engineer and then chief forester of the Forest Service, tried to slow the harvest, but he ran into an angry U.S. Representative Les AuCoin of Oregon. AuCoin, heavily backed by timber industry political action money, challenged Peterson in a 1986 appropriations subcommittee hearing. AuCoin's shortsightedness and the pressure he brought to bear were apparent in an exchange captured by the Portland *Oregonian* in a series of articles it published on the ancient forests in 1990. He asked why the chief was asking for less money for timber sales in 1987 than had been spent in 1986. Peterson, who had been trying to scale down the harvest to spread it into the next century, said, "When you are looking to balance the budget this year, the problems you may have after the year 2000 don't loom as large."

"You won't be the chief in the year 2000. I won't be in Congress . . ." AuCoin replied. "We don't have to worry about that."

"I might still be around, I don't know," the chief joked.

"Chief, if the second half of the fiscal year volume flow in Region 6 is as depressed as the first half, I am not sure I would make that statement," AuCoin said.

AuCoin and Hatfield successfully added enough money to the 1987 budget to force the agency to cut an additional billion board feet of timber in the Northwest. They added 300 million board feet to the Forest Service's timber target in 1988 and 200 million board feet in 1989.

The recovery plan for the northern spotted owl projects an overall harvest in Oregon and Washington of 1.27 billion board feet annually by 1995, a far cry from the annual harvests of 5 billion board feet of the late 1980s. In 1989, there were 347 sawmills and more than 110,000 people working in the wood-products industry in Washington and Oregon. The recovery plan estimates that the number of jobs in the industry will drop by nearly a third, or 32,000 workers, because of efforts to protect the owls. Many of those woodworkers might still have had jobs into the late 1990s if not for the enormous harvests of the late 1980s.

At the same time that scientists were warning foresters about the threat to the northern spotted owl, similar messages were being sent from fisheries biologists about the Snake River salmon. The completion of the Lower Granite Dam on the Snake River in 1975 was the final straw for salmon that spawned in Idaho. The last of eight dams on the Columbia and Snake rivers between Idaho and the Pacific proved too much for many of the fish, especially on the migration back to sea; fully half of the fish leaving Idaho never made it though the long, slow-flowing reservoir to Lower Granite. The juvenile salmon were either lost in the deep, still water or eaten by predators. In the drought year of 1977, scientists said 95 percent of the juveniles died before reaching the ocean.

In January 1978, Pocatello, Idaho, attorney Herman McDevitt wrote two letters to the regional director of the U.S. Fish and Wildlife Service demanding that Snake River spring and summer chinook stocks be considered for protection under the federal Endangered Species Act. McDevitt, a member of the Pacific Fishery Management

Council, said that scientific evidence "seems to predict an exceptionally weak run of these fish for 1978 and subsequent years."

His letters were not official petitions for listing, an important dis-   121
tinction because the National Marine Fisheries Service, the agency that decided to conduct the status review, was not forced to go through the formal review process required once a petition is received, with all of its deadlines and public involvement. Instead, the agency decided in October of that year to review all salmon and steelhead populations in the upper Columbia River Basin, a far easier process, one that could take as long as the NMFS wanted.

"I remember Herm indicating there was an enormous amount of resistance to him filing a petition," says Ed Chaney of Eagle, Idaho, executive director of Save Our Wild Salmon, a group working to protect salmon. "He elicited certain promises that they would move forward with the review if he didn't file a formal petition."

In 1981, the NMFS suspended the Endangered Species Act review, reaching no conclusion on the status of the upriver stocks. The decision was made, according to the unpublished Federal Register announce-ment, because of fisheries protection provided in the recently approved Northwest Power Act and new conservation efforts by the states and the tribes due to treaty-rights court decisions. The Northwest Power Act in particular appeared to offer strong protection for salmon by promising "equitable treatment" for fish and power. That same year only twenty-six sockeye salmon returned to Redfish Lake in Idaho.

"Everyone was enormously optimistic about the power act," says Lorraine Bodi, who was an attorney for the NMFS in 1981. "We thought it had the clearest language in the world. We thought we didn't need the Endangered Species Act."

Even environmentalists saw no need to proceed with a salmon listing. "There was a sense of euphoria when the power act passed," says Ed Chaney. "We thought we had won. We all bought off on it."

Although it had suspended the Endangered Species Act review of salmon stocks, the NMFS was not ready to formally terminate its endangered species review. By law it was supposed to publish a formal statement of termination or suspension in the Federal Register. A statement of suspension, which left open the possibility of reopening the review later, was prepared, and the NMFS's uncertainty was clear in the unpublished document. "These data have not been analyzed based upon legal or policy considerations of the ESA, and NMFS and

FWS have not made a determination on whether any of these populations should be listed as endangered or threatened species," the un-

signed Federal Register notice said.

It was unsigned because the Office of Management and Budget (OMB) rejected it and asked instead that the salmon status review be terminated, essentially stating that the fish were not threatened or endangered. The OMB under the Reagan administration had become increasingly meddlesome in the affairs of virtually all federal agencies. It went beyond its traditional budgetary role. The status review was just another example of its involvement in resource policy-making.

In a memo dated February 1, 1982, Dale Evans, the biologist in charge of the review, told NMFS Regional Director H. A. Larkins he recommended the salmon endangered species review be placed in inactive status. "At this point we believe that termination may be inadvisable. It could create the overall impression that the status of the imperiled populations is not so severe as to warrant ESA protection," Evans wrote. "Since at this point, the remedial successes are largely projections, we believe that ESA protection is still a possible option and the review should be inactivated rather than terminated."

Buried within the National Oceanic and Atmospheric Administration, which itself is a part of the Department of Commerce, the National Marine Fisheries Service prides itself on having a higher percentage of Ph.D.s on its staff than any other governmental agency. Since Commerce is not otherwise directly charged to manage fish, wildlife or natural resources in general, the agency finds itself with little political clout in interagency politics, such as budgeting. So it is not surprising that the NMFS has shown itself to have even less enthusiasm than the FWS when it comes to listing endangered species.

The NMFS's job is to protect and manage marine fish and wildlife, while the FWS protects and manages fish and wildlife inland. They were placed in different federal agencies partly because management of marine resources began as trade activities between the United States and other countries that use ocean resources. When the Endangered Species Act was passed, the two agencies each were given authority to list and protect endangered species. The NMFS began developing its own candidate species list in 1989. Far smaller than the FWS, the NMFS has had limited resources to carry out its own endangered species program.

The NMFS has listed twenty-three species—six since 1976. Of the twenty-three, no marine plants or invertebrates have been listed. Only three of the species on the list are fish, which critics say demonstrates   123 that the commercial fishing industry, the major public-interest group with which the agency deals, exerts political influence on its listing program. "Unfortunately, the agency's overall attitude has been to avoid listing species, say many NMFS officials, because listed species might cause conflicts with the fishing industry," wrote Amos Enos of the National Fish and Wildlife Foundation in a 1990 study of NMFS's effectiveness.

The foundation's study blamed inconsistency in leadership for the agency's poor performance in endangered species management. "One administrator emphasizes protected species, the next does not," Enos wrote. "A major complaint of those NMFS officials who want the program to do well is the lack of interest and positive support from the top, not to mention occasions when the [Commerce] secretary or [National Oceanic and Atmospheric Administration] administrator actually works against the program."

Charles Karnella, NMFS chief of the protected species division, says he did not see such high-level interference. "As far as I can recall, that never has happened," he says. As for the other criticisms in Enos's report, Karnella said many of the recommendations have been accepted, including the hiring of listing and recovery coordinators. "I think the people in the agency believe the agency has taken several steps to beef up the protected species program."

Karnella's optimism for improvement has not been borne out in the decisions of the late 1980s and early 1990s. If science has been the overriding factor in decision-making, then the NMFS must believe that fish can swim up dry channels and dams are hardly a threat to their survival. The problems the listing process poses for the NMFS are best demonstrated by the case of the winter-run chinook salmon in the Sacramento River.

In 1985, the American Fisheries Society (AFS) petitioned the NMFS to list the winter-run salmon, based on the decline of the run in the northern California river to about 2000 per year, down from about 84,000 in the late 1960s. This is about a 97 percent decline.

Studies showed that the Red Bluff diversion dam, operated by the Bureau of Reclamation to provide irrigation water to farmers, was

blocking fish passage by drying up channels and raising water temperatures. Other problems included mine-runoff pollution and other pri-
124   vate diversions of water from the river.
     All young fish hatched in 1976 and 1977 had been killed by drought-
induced rises in river temperatures, made worst by the Red Bluff
diversion. In 1980, the run had dropped to 1156 fish. By 1984, it was
clear to scientists that the run, which is the only winter run of salmon
on the entire Pacific Coast, was on the brink of extinction, says Cay
Goude, a California biologist and former president of the western
division of the American Fisheries Society. "But NMFS did nothing,"
says Goude.
     After the AFS petition, the NMFS decided in 1987 not to list the
species as threatened, depending instead on a voluntary ten-point
cooperative restoration plan offered by state and federal authorities. In
1989, the run size for the year plummeted to 533 fish, and finally the
NMFS acted. It published an emergency rule that declared the species
threatened and forced the Bureau of Reclamation and the state of
California, the two entities that controlled the Red Bluff diversion, to
provide additional water and passage for the now seriously depressed
fish stocks. In October 1989, it permanently listed the winter run as
threatened, but it may have been too late. The numbers today are so
low that the genetic diversity necessary for long-term survival already
may have been lost.
     When the NMFS decided to list Snake River sockeye in 1991, the
decision was more clear-cut and scientifically based. In contrast with
its 1981 review and 1989 winter-run listing decisions, the NMFS did
not consider the promises of anyone to protect sockeye. There were
not enough sockeye left to promise to protect. Instead, NMFS scientists had to decide just what constituted a sockeye. The argument
revolved around the building and destruction of Sunbeam Dam on the
Salmon River. Between the time it was built in 1911 and the time it was
breached in 1931, there is uncertainty about whether sockeye continued to scale the dam's poor fish ladders. What all scientists agree upon
is that sockeye rebounded after the dam was removed and didn't begin
to disappear until the late 1950s. Scientists are divided over how the
run was restored after 1931. Was it by descendants of a few remnant
native sockeye that managed to scale the inadequate fish ladders while
the dam stood? Or by some resident Redfish Lake kokanee—the
landlocked, freshwater form of sockeye—drifting seaward after the

dam was breached, somehow realizing their genetic potential to become sockeye salmon?

Experts waged the argument within the technical committees of the NMFS's endangered species review. Don Chapman, a fisheries biologist under contract to the Pacific Northwest Utilities Conference Committee, said the original ocean-run sockeye was eliminated by Sunbeam Dam and that restoration came solely from Redfish Lake kokanee.

"The Redfish [Lake] kokanee population is the same population unit that supports the 'sockeye' population of Redfish Lake," Chapman wrote in a report submitted to the agency. "We do not find evidence to indicate that the kokanee of Redfish Lake fit the endangered or threatened criteria." Chapman's clients—nearly every Pacific Northwest electric utility and many large electricity users—wanted to avoid endangered species listing.

Susan Broderick, the Shoshone-Bannocks' fisheries coordinator who wrote the petition asking that sockeye be listed, agreed that there is strong evidence the two forms interbreed and that the kokanee is able to transform into the sockeye. That ability is integral to the tribes' plan to restore the run. But Broderick also says that both the kokanee and post-Sunbeam sockeye show unique genetic characteristics of the original population, such as lake shoal spawning.

Broderick says there can be no Snake River sockeye unless the fish make the trip to the ocean. "I want our sockeye run back even if it's kokanee," she argues. "You've got a gene pool able to migrate 900 miles if we give it a chance."

The NMFS finally did list sockeye as endangered but made a distinction between sockeye and what it called "sea-run kokanee." It found that there were measurable genetic differences between the two, and it wanted to protect only the genetic stock of the sockeye. Such a ruling had significant management implications. It severely limited the spawning stock in Redfish Lake that was protected to only those fish clearly determined to be sockeye. Any sockeye that might stray to another lake to spawn might not be considered protected. If at some time in the future it was determined that the sockeye was extinct, returning "sea-run kokanee" would have no protection. It made recovery even harder and extinction more likely.

The listing decision for Snake River chinook was even more controversial. The NMFS listed fall chinook as threatened, which means under the law that a species may soon become endangered if something isn't done. In 1991, only seventy-eight fall chinook returned to spawn in the Snake River. Chinook may not soon become endangered, they may soon become extinct. But listing them as threatened meant there was more flexibility for allowing fish to be taken incidentally by fishermen or by the Army Corps of Engineers in their dams or in the fish transportation program.

With each decision, the NMFS has become more adept at organizing and responding to the arguments of the various interest groups. And science, not promises of protection, is driving the listing process. The main reason for this is public scrutiny. With the future of the entire Pacific Northwest on the line, the NMFS can't afford to make the same mistakes it has made in the past.

The endangered species reviews of the spotted owl and Snake River salmon attracted much attention in the late 1980s, but they were the exceptions rather than the rule. Many listing reviews have gone unnoticed, except for the direct involvement of a few activists. Hundreds of other species remain in listing limbo, either lost in the middle of a process never completed or moved to candidate species designation, which the agency says means they warrant immediate listing and protection. One of the species caught in this bureaucratic purgatory is the Bruneau Hot Springs snail, a tiny mollusk barely bigger than a ball point on a pen.

The Bruneau Hot Springs snail is one of ninety snail species that once lived in the ancient glacial Lake Idaho, which spread 300 miles across southern Idaho, south of present-day Twin Falls. About 12,000 years ago the lake drained when a natural dam washed out. Today, only thirty snail species remain spread around in islands of livable habitat. Many of the species have very specific requirements for life, such as clean, clear water. The Bruneau Hot Springs snail is unique because it can survive in water as warm as ninety-six degrees Fahrenheit. It is found only in hot-water springs along the Bruneau River and its tributaries. The Bruneau River is a tributary of the Snake River located about 120 miles southeast of Boise, Idaho.

The snail's existence depends on an aquifer, fossil groundwater perhaps left over from Lake Idaho, that underlies its habitat. It was discovered in 1952 in a hot springs known as Indian Bathtub, which had been used for bathing by Indians for centuries. The snail population has been studied extensively since its discovery and was known to be stable until the early 1960s.

In the late 1960s, ranchers and farmers in the area began pumping millions of gallons of water out of the aquifer to irrigate alfalfa to feed their cattle. Today, as many as 18,000 acres of farmland are irrigated with water pumped from the aquifer. Water scientists say the farms are pumping far more water than the aquifer is getting back in recharge. The evidence shows at Indian Bathtub. In 1954, some 2200 gallons of water per minute bubbled from the springs into the tub. In 1972, the flow had dropped to 458 gallons per minute. The flow dropped to 140 gallons per minute by 1978, and by 1985 the springs stopped flowing altogether in July and August.

D. W. Taylor, the biologist who until recently had done the most thorough studies of the snails, documented the snail's disappearance as the flow dropped. By 1980, Taylor said, the loss of the snail habitat in the springs was extensive, and in 1982 he recommended protection for the snail under the Endangered Species Act.

In 1985, the Fish and Wildlife Service finally proposed listing the snail. Ranchers in the area immediately were alarmed because they saw the possibility that their irrigation pumping would be limited if the snail was protected. Such limits would cut directly into their income and might put some of them out of business. Other farmers saw it as precedent-setting, leading to a larger water grab for a variety of fish and wildlife programs. Through the Idaho Farm Bureau, they pressured the FWS not to list. Following the comment period, despite clear evidence, the FWS dropped the snail from its list of candidate species, but made no formal decision.

The smell of politics hung over the decision like a cloud. In 1988, the state's two most powerful men in Washington, in a rare case of leaving a paper trail, documented the manipulation of the listing process. Republican Senators James McClure and Steven Symms were among the strongest guardians in Congress for western farmers, loggers, miners and others whose livelihoods depended on public lands and resources. McClure, an attorney and former chairman of the Senate Energy and Natural Resources Committee, had become the leading spokesperson in Congress against more intrusive federal control of the West's land and water. Symms, a former fruit farmer, was a Sagebrush Rebel and an unbending conservative who believed fiercely that private property rights were inviolate. The Idaho senators left no doubt what would happen if former FWS director Frank Dunkle listed the snail as endangered. "Without your assistance, the need to preserve Idaho livelihoods will require us to address the issue legislatively," McClure and Symms wrote in a letter dated February 23, 1988.

Congress was going through one of its periodic reauthorizations of the Endangered Species Act, when the actual funding for the act is renewed for a period of five years. The Idaho senators, in the letter, threatened to hold up reauthorization of the funding of the act until Dunkle responded to their "request." McClure, the ranking minority member of the Senate Energy and Natural Resources Appropriations subcommittee, and Symms, a member of the Environment and Public Works Committee, were well placed to carry out their threat to hold up or cut funding to the endangered species program if Dunkle didn't cooperate. Dunkle didn't wait long. On April 6, 1988, he told McClure he would comply. "At your request, the service will delay its final decision on listing the Bruneau Hot Springs snail pending adequate funding and cooperating agencies' approval and implementation of the management plan," Dunkle wrote. "However, if during this time period populations deteriorate to unacceptable levels, the service will take emergency actions to prevent extinction from occurring."

In return for holding off listing, McClure placed $850,000 funding for research into Interior appropriations bills. He said that by conducting the research and management, the FWS could better protect the snail by building consensus with the farmers who own both the snail's habitat and, under Idaho law, the water. As often was the case with McClure's manipulations of the appropriations process, there was some good done. The money did fund research that found several more sites where the snails survive along Hot Creek and the Bruneau River. The U.S. Geological Survey has since mapped the aquifer, and fences were built to stop livestock from walking through the delicate snail's home.

However, the major threat to the existence of the snail remains groundwater pumping. Under Idaho law, pumping more water from an aquifer than is recharged is illegal, and the state Water Resources Department is supposed to designate areas like Indian Bathtub critical groundwater areas, where no future water development is permitted. No such action has been taken.

Dunkle defended the decision, saying that taking only science into account when making listing decisions makes no practical sense.

"There are those in the service who think the only thing they have to do to save a species is to list it," he said. "We have to use rational approaches."

However, a 1982 amendment to the Endangered Species Act had made clear that listing decisions must be made "solely" on statistical data available, making Dunkle's "rational approaches" illegal.

In 1992, six years after the FWS was supposed to have ruled on the snail listing, environmentalists filed a notice of intent to sue. The Land and Water (LAW) Fund of the Rockies, a nonprofit regional environmental law center based in Boulder, Colorado, filed the suit on behalf of the Idaho Conservation League and the Committee for Idaho's High Desert. The environmentalists said the handling of the snail was a clear violation of the Endangered Species Act.

"All our clients are asking is that the Fish and Wildlife Service comply with the nation's environmental laws," said Kate Zimmerman, LAW Fund senior attorney, in a press release. "They are six years too late with the decision on the Bruneau snail. There is a huge backlog of species facing extinction, waiting for the FWS to act. Our clients question the commitment of the FWS to uphold the Endangered Species Act."

The environmentalists' suit worked, and in November 1992, the FWS agreed to rule on the snail's status by January 15, 1993. It listed the Bruneau Hot Springs snail as endangered in January 1993.

It is true that there are 600 species listed as candidate species that the FWS says are imperiled enough to warrant immediate listing and protection under the act. Another 3000 species are designated as "C-2" species, or lower-priority species, of which up to 1800 eventually will qualify for listing. At current rates of listing, dramatically higher than during the Reagan years, the agency says it will take from thirty-eight to forty-eight years to list all that qualified for listing in 1990.

For the most part, officials in the Bush administration took a more scientific approach. When John Turner, former president of the Wyoming Senate, took over as director of the FWS in 1989, he gave his field offices the green light to speed up the listing process nationwide. More important, Turner made it clear that only biological factors would be considered in the listing process, as the law requires. "I do believe the credibility the people of this country have in my agency is that we follow the law, and the law says you use good science," Turner said.

Turner was the point man for the Bush administration's endangered species program. He was thrust into the national limelight in June 1990, when he issued the decision to list the northern spotted owl as threatened. But Turner found his popularity among environmentalists and congressional Democrats a political problem both at home, where many said he hoped to return to run for governor, and in Washington, where he had quietly carved out a niche as one of the conservation consciences of the Bush administration. With the support of Bush, whom he once guided down the Snake River, Turner restored a measure of respectability to his agency. Yet while the FWS had begun complying with the listing provision of the act more responsibly, it continued to face a bevy of lawsuits concerning the designation of critical habitat and other enforcement provisions of the act. And listing indecisions of the past, such as the Bruneau Hot Springs snail, continued to haunt the agency.

Turner's credentials fit the political requirements of FWS directors in a Republican administration. He came from a western state dominated by resource issues. For the most part he used his smooth political savvy, honed in twenty years of politics in Wyoming, to negotiate around the land mines of Washington, D.C.

Turner also benefited from his experience as a field biologist and from growing up in the wildlife-rich Yellowstone region. He did some of the pioneer research on eagles in the Yellowstone area in the 1960s and also had surveyed trumpeter swans and written papers on grizzly bear management. So in 1990, when he made the decision on whether to list Rocky Mountain trumpeter swans as endangered, he didn't just rely on experts. He personally called biologists he knew in the field to get their opinions. "The advice I got from those experts was that it doesn't warrant listing at this time, but you had better keep a close watch on them," Turner says.

Turner's decision on the trumpeter, like the decisions of the NMFS on salmon, was more closely tied to science and thus legally defensible. The trumpeter decision showed the complexity of making any listing decision that affects politically powerful groups, such as Idaho irrigators.

After more than fifty trumpeter swans died because of extreme cold and low water flows on the Henry's Fork of the Snake River in 1989, the Idaho Wildlife Society petitioned the FWS to list the Rocky Mountain population of the magnificent migratory birds as threatened. Regional FWS offices in Denver and Portland recommended that the agency accept the petition to study listing for the Rocky Mountain population. But in December 1989, Richard Smith, assistant director of the agency, sent the proposal back to the regional offices for additional discussion.

131

When Smith issued the decision, regional staff were negotiating with the Bureau of Reclamation and the Committee of Nine, which represents eastern Idaho irrigators, in an attempt to get assurances that water would be released from Island Park Reservoir in the winter to prevent another loss of trumpeters. But an internal FWS memo, written one day after Smith's letter rejecting the petition, said the decision would undermine water talks. "Personnel in Boise indicate that both groups [bureau and water users] are 'dragging their feet' since the 90-day finding has not been signed, and the threat of listing is not ominous," the memo stated.

No agreement was ever reached, and the agency officially rejected the petition in April, based on its own commitment to provide funding to purchase water for eastern Idaho's Henry's Fork swans, and on a program to expand the swans' winter range by physically moving them from Harriman State Park. But FWS memos reveal that the agency received no assurances that it could purchase enough water in extremely dry winters, such as that of 1988–89. As with the salmon in 1981 and the winter run salmon in 1989, there were no guarantees.

Instead of listing, the agency, along with Idaho and Wyoming wildlife officials, embarked on an aggressive and apparently successful program to transplant the Henry's Fork swans to long-abandoned habitat throughout the region. Chuck Peck, who coordinated the swan transplant for the FWS, says it was not done in response to the unsuccessful listing petition. The agencies have recognized a need to expand the trumpeters' range for at least a decade, Peck says. Many of the swans that crowd the winter habitat come from western Canada in growing numbers every year.

But as with the Bruneau Hot Springs snail, research and management money for trumpeters came flowing in after the decision not to list. The transplant program has been largely successful so far, making the case for Turner's strategy. But trumpeters have protection under federal waterfowl laws that most endangered species don't share.

Turner would have liked to have used the same strategy that was successful for trumpeters for protecting other species teetering on the brink of listing. But by the time he was heading the agency, there were few opportunities like the trumpeter left. The delays and inaction of his predecessors left the tough calls to him. In the case of the northern spotted owl, the scientific evidence was clear. Yet Turner agonized over the decision just the same.

"I'm particularly sensitive to that, coming from Wyoming, where you know what tough times are," Turner says. "I went to the Northwest. I met the young families that just want to pay their mortgages, keep their kids in schools and love the woods and love their work. They just want to be assured they'll have a job."

In 1990, Turner's first full year as director, 47 species—37 in the United States—were listed, about average for the agency in the five preceding years. More indicative of Turner's initiative was that 106 species were proposed for listing, including 89 in the United States. These figures still leave the agency with a tremendous backlog. "I hope to bring some expediency to that listing process," said Turner at that time. "We are behind in it and we are justly criticized for it."

Despite Turner's intentions, the political assault on the listing process was not over. In 1991, pressure from the Idaho Farm Bureau delayed the listing of five mollusks that live in the Snake River and whose listing threatens irrigation and power dams. Four species of snail—the Bliss Rapids snail, Utah valvata snail, Snake River physa snail and Idaho spring snail—and the Banbury Springs limpet were proposed for listing as endangered species by the FWS in December 1990. The mollusks are believed by biologists to live only in a few places along a forty-mile stretch of the Thousand Springs area of the Snake River and perhaps upriver near Pocatello, Idaho. Like the Bruneau Hot Springs snail, the five mollusks are the remnants of mollusk species that lived in Lake Idaho.

The decision to list was based in part on the snail studies conducted in 1982 by D.W. Taylor. Peter Bowler, a California biologist who grew up in the area near the mollusks' habitat, and Terry Frest, a malocologist, or mollusk biologist, from Spokane, Washington, also had extensively studied the mollusks and concluded endangered species protection was warranted. The actual proposal to list came long after the deadlines mandated by the act had passed, leaving developers with the opportunity to move forward with plans for dams, fish farms and other activities threatening snail habitat. Senator McClure, in his final days in the Senate, had kept the agency from proposing listing with veiled threats and demands for additional studies. After the decision to propose listing was released, the Idaho Farm Bureau began a campaign to get Jay Gore, the FWS biologist involved in the spotted owl listing controversy of 1987, fired.

During the mollusk review, a scientist on contract for several utilities pushing hydroelectric projects along the middle Snake River, where the mollusks lived, had offered new information that at least two of the snail species—the Bliss Rapids snail and the Utah valvata snail—were found in a stretch of the river above American Falls Reservoir, about 100 miles east of the main mollusk habitat. Richard Konopacky, of Meridian, Idaho, said he was able to find the snails in just three short sampling trips. He said his finding showed that the snails might be far more numerous than originally thought.

Gore, who was in charge of the mollusk status review, said Konopacky's information was not compiled in a form that was usable for the original proposal documentation. He told Konopacky to complete his research and put it in final form for presentation in the public comment period following the proposed listing.

The proposal and Gore's handling of it came under attack by the Idaho Farm Bureau in April 1991. Mike Tracy and Rayola Jacobson, two of the bureau's lobbyists, blasted Gore in the press for leaving Konopacky's research out of the decision documents. Tracy questioned the credentials of Bowler and Frest and said Gore should be fired.

The pressure appeared to have the desired effects. Tracy met with Gore's boss, Charles Lobdell, FWS field supervisor, in Boise on April 12. In an April 23 memo to Lobdell, Gore related the fallout of the meeting: "Shortly after the meeting you came to me and commented that you were coming under intense pressure on the snail listing and that you and I may lose our jobs over this one.

"On April 15," Gore continued, "I received information that Ms. Rayola Jacobson, Idaho Farm Bureau, had stated that she wanted me removed. The Farm Bureau had even contacted a Senator's office to discuss my removal, for my involvement with wolf recovery and the five snail listing. On April 16 you again stated to me the pressure you were feeling on the snail listing and that you would probably pull this listing."

The days of Gore's career in the Fish and Wildlife Service were numbered. Lobdell gave Gore a negative evaluation based primarily on his decision not to include Konopacky's raw data in the proposal package. At his own request Gore was taken off the mollusk listing team. When he agreed to turn over to another staffer the writing of the final rule to list the mollusks, Lobdell wrote, "I will reassign the final rule (if we do one)." Gore, already frustrated by the lack of funding for wolf recovery efforts in Idaho and wounded by the spotted owl controversy, looked for a new job. Later in the year he went to work in Washington, D.C., for the Forest Service, overseeing research and management of the northern spotted owl.

Meanwhile, Lobdell met with the Idaho Farm Bureau, Idaho Waterusers Association, Idaho Power Company and Konopacky to hear their criticisms of the process. He agreed to enter into a cooperative financial arrangement with the groups to collect additional information. It included convening a panel of scientists to review the listing proposal. Bowler and Frest, the two most recognized experts available, refused to participate in the panel discussions led by C. Michael Falter of the University of Idaho's Department of Fish and Wildlife Resources because it was rigged against them. Including Falter, the panel had Konopacky, Lobdell, a biologist for Clear Springs Trout Company (a hatchery near snail habitat), an Idaho State University scientist, Gore's successor and another University of Idaho biologist. Following the panel discussion, Falter, who was in charge of summarizing the discussions, consulted with Bowler, Frest and Robert Hershler, a mollusk expert from the Smithsonian Institution. In his report, issued March 24, 1992, Falter supported the recommendation to list the five mollusks. "The possibility of populations increasing or additional discoveries of these [mollusks] in poorly sampled or unsampled habitat is remote considering the rapid habitat deterioration," he wrote.

Konopacky and Tracy were furious. They had successfully stacked the deck in their favor and still Frest and Bowler's findings were given more weight. "It's a biased, slanted report and has nothing to do with science," Tracy said angrily after the report was released.

Just as they had done with the Bruneau Hot Springs snail, the Idaho Conservation League, the Committee for Idaho's High Desert and the Land and Water Fund of the Rockies filed a notice in July 1992 of their intent to sue the FWS for not listing the mollusks by the deadline. In October 1992 they filed the suit in federal court.

Finally, in February 1993, the mollusks were listed. Four were listed as endangered, with the Bliss Rapids snail listed as threatened, in part because of Konopacky's research.

Forcing the agency's hand in court had worked before to help the northern spotted owl and was employed successfully to force the FWS to list the marbled murrelet, another bird that depends on the old-growth forests west of the Cascades. The murrelet is a medium-size seabird that spends most of its life on the open ocean feeding on a variety of aquatic species. It nests as far as fifty miles inland in mossy, old-growth Douglas fir trees. Scientists have had a hard time determining just how many murrelets are left along the shores of Oregon, Washington and California. Biologists could find only twenty-six nests, all in large, old-growth trees. The liquidation of old-growth habitat along the coast was leading most scientists to say that murrelet numbers were declining significantly. Some reported large numbers of birds returning from the sea only to find traditional nesting areas clearcut. For these flocks it was like waking up after a nuclear war to find the world destroyed.

The murrelet was proposed for listing as a threatened species in June 1991. The FWS did not decide by June 1992 to list, proposing instead to delay the decision for six months. The Sierra Club Legal Defense Fund sued the FWS in federal court, demanding that it immediately list the murrelet. In September 1992, a federal judge ordered the agency to list the bird as threatened, and the agency complied.

It took environmentalists a decade to hit their stride as they successfully used the courts to force hostile administrations to carry out the mandate of the Endangered Species Act. The election of President Bill Clinton, a man who says he supports the Endangered Species Act, might be expected to end the need for court intervention for environmentalists. Clinton appointed former Arizona governor Bruce Babbitt as Interior secretary, placing a man with solid environmental credentials at the central seat of natural resources power. Babbitt, the former president of the League of Conservation Voters, expressed strong support for carrying out the Endangered Species Act. But he also expressed support for Turner's basic strategy: to avoid polarizing listing controversies by saving species before they must be listed. A major critic of Bush administration policies on endangered species, George Frampton, former president of The Wilderness Society, was appointed assistant Interior secretary for fish, wildlife and parks, which is the post in charge of the U.S. Fish and Wildlife Service and the National Park Service. Along with other former environmental activists peppered throughout the new Clinton administration—including Vice President Al Gore—the policy on endangered species could be the most obvious change brought about by the 1992 election.

136

Now the Idaho Farm Bureau and other strident Endangered Species Act critics are shut out of the new administration in ways that environmentalists were in the Bush and Reagan years. Faced with an unfriendly administration, the Farm Bureau has taken a page from the environmentalists' strategy book, taking the FWS to court in an attempt to overturn the Bruneau Hot Springs snail and mollusks' listing decisions.

But despite the strong environmental representation in the new administration, most industry groups have taken a wait-and-see approach to Clinton's resource policy team. Clinton had promised in his campaign to convene a forest summit in the Pacific Northwest to resolve the old-growth forest debate—a promise he kept on April 2, 1993—in a way that preserves the environment and provides jobs to the people in the region's rural communities.

Clinton takes seriously Gore's view that protecting the environment and enhancing the economy can go together, and many believe he will prove it in the Pacific Northwest. His main obstacle will be the unwillingness of radical voices on both sides of the argument to accept a new reality, a compromise that forces both sides to move away from old, comfortable debating ground.

For industry groups it means allowing the transition to a new economy that is less dependent on natural resource development. For environmentalists it may mean allowing new, streamlined regulatory systems and market-based incentives to replace burdensome environmental impact reviews and appeals processes. Any successful strategy will be based on the goal of preserving and restoring the natural ecosystems on which fish, wildlife and humans all depend.

6. The Sword and the Shield

THE kind of jobs-versus-owl, development-versus-preservation confrontation that has taken place over the northern spotted owl has happened only rarely since the Endangered Species Act was passed in 1973.

The case of the owl and its effects on the coastal timber industry is as significant for its rarity as it is for its potential effects. Even though the listing of a species gets most of the attention, listing in itself would mean very little if not for the language in Section 7 of the act. Section 7 requires federal agencies to take "such action necessary to ensure that actions authorized, funded or carried out by them do not jeopardize the continued existence of an endangered species." Simply, the federal government is required to use its vast power to protect species rather than to destroy them.

This is absolute. No equivocation. Nowhere else in this law or any federal law has wording been so clear or so protected from legal misinterpretation. Yet while the legal mandate has been clear, the actual implementation has been murkier than mining runoff.

Few believe Congress or President Richard Nixon fully realized what they were approving when they passed and signed the Endangered Species Act and this one provision. It is doubtful whether Section 7's absolute mandate could be included in environmental legislation today. Senator Mark Hatfield (R-Oregon), who helped write the act, said it was meant to be used as a shield instead of a sword.

Even though the endangered species sword is sharp, it is rarely raised by the government in defense of its kingdom. Instead, the two agencies that carry it—the U.S. Fish and Wildlife Service (FWS) and

137

the National Marine Fisheries Service (NMFS)—enforce their will with a quick glance at their scabbards. It has been environmentalists,

through the courts, who have wielded the rapier with the most skill. As far as the agencies have been concerned, even when the battle is joined, the most that can be expected is a little saber rattling. That's because the act's most effective power is exercised by formal and informal consultations between the FWS or the NMFS and other federal agencies concerning proposed activities. Often these consultations take place behind a shield of secrecy.

Officials decide during consultations if projects would harm endangered species. Instead of stopping projects, consultations usually bring about agreements to change them so that their effects on endangered species are reduced or eliminated. Take, for instance, a timber sale. A U.S. Forest Service biologist analyzes the sale that is laid out by a forester, first going down the list of endangered species that may be affected by the harvest and then deciding whether the sale needs to be altered. He or she makes recommendations and then issues a decision that it won't affect, may affect or will affect endangered species. If the decision is that it will affect, then the forester probably will be sent back to the computer terminal to restructure the sale. If the decision is that the sale will have no affect or that it may affect an endangered species, then the biologist calls the FWS and enters into what is called informal consultation. Now two biologists are talking about the ramifications of the sale. If it is a "no effect" decision, the biologists may or may not go out on site for an afternoon. In either case, they then exchange letters to provide proper documentation, and the FWS is considered to have officially concurred with the decision.

If the Forest Service biologist issues a "may affect" decision—and often in the 1980s he or she did so over the frowns of Forest Service supervisors—then the two biologists will look for ways to change the sale to avoid the effects—for example, move a road so that it doesn't threaten an eagle's nest, or leave strips of timber for escape cover for grizzly bears. By this time, the district forest ranger, the Forest Service biologist's immediate supervisor, is inevitably involved, since negotiations are taking place. The supervisor wants to make sure that the measures taken to protect the endangered species trim the least amount of timber from the sale as possible. It is Forest Service policy to avoid going to formal consultations whenever possible, so depending on how much pressure the FWS applies, the supervisor may make all the changes the biologists say are needed.

If the FWS decides to take the issue to formal consultation, then its 139 biologists write a biological opinion. This is where its power and the power of the Endangered Species Act lie. Now, legally, it is the FWS in control. It must decide, in the biological opinion, whether the federal action, the timber sale, jeopardizes the existence of any endangered species. If the FWS issues a jeopardy opinion, then the trees can't be cut, or the roads can't be built. But the FWS also must provide "reasonable and prudent" alternatives under the law. So it might include an alternative with dramatically scaled back logging or mitigation measures to offset the effects of the timber harvest.

The overwhelming majority of consultations are informal. Biological opinions are as infrequent as caribou sightings in the Lower 48 states. Jeopardy opinions are as rare as Snake River sockeye. The Fish and Wildlife Service conducted 28,000 informal consultations in 1990, and only about 600 to 700 went to formal consultation. Less than 1 percent resulted in jeopardy opinions, and only two halted projects: a proposed federally funded water project in Colorado and a harbor project near San Francisco. In Region 1, which includes Idaho, Oregon, Washington, California, Nevada and Hawaii, FWS staff conducted 4141 informal consultations and 204 formal consultations, issuing one jeopardy opinion, in 1990. In the last ten years, only three jeopardy opinions have been issued in Idaho, thirteen in western Montana and five in western Wyoming. From 1987 through 1991, the FWS consulted with federal agencies on endangered species 75,000 times, and only nineteen times in those five years were projects ultimately canceled.

Many environmentalists point to these statistics to show that the act is not working because it is allowing too much development to slowly nibble away at the habitat and resources needed to prevent extinction. The informal consultation particularly worries many observers. What worries many environmentalists is that the informal consultation takes place out of the public limelight. It goes on behind closed doors, making the negotiations between the federal agencies involved hard to scrutinize.

There is little opportunity for public involvement, so the judgment of a small circle of biologists is all that is weighed. Moreover, the informal process allows a less structured examination of the cumulative effects of

140

activities in an area. One timber sale and one road might not threaten the existence of a species, but ten timber sales and fifty miles of road spread out over a larger area may make the entire area worthless as an endangered species habitat.

Yet others argue that the act wasn't meant to halt development except in extreme cases, such as is necessary to protect the northern spotted owl. Hank Fischer of Missoula, Montana, has been regional director of one of the Endangered Species Act's most vocal supporters, the Defenders of Wildlife. He says environmentalists are wrong when they say the act is ineffective because it has not produced more clear victories for species. "The strength of the Endangered Species Act has always been in modifying developmental activities, not in stopping them," Fischer says. "And to the extent polls show that moderating negative development is precisely what the public supports, the law has been highly successful."

Steven Yaffee, a professor at the University of Michigan and author of the book *Prohibitive Policy: Implementing the Federal Endangered Species Act*, describes the consultation process as a kind of political pressure valve that balances the need to protect endangered species with the need to protect the political support of the law itself. With all of its flaws, the consultation process, with its mandate against jeopardizing a listed species, has survived in part because of its flexibility.

"It survives because it makes sense substantively, it has been implemented flexibly and adaptively (perhaps too much at times), and environmentalists have mustered enough support in Washington to turn away major challenges," Yaffee wrote.

Industry critics of the Endangered Species Act say that even if the consultation process rarely stops a project legally by itself, it results in long delays that add to the costs of projects. This charge was specifically disputed in a 1987 investigation conducted by the General Accounting Office, which examined the impact of the consultation process on water projects in seventeen western states.

"Consultations carried out under the Endangered Species Act have had little effect on western water projects," the report's authors concluded. "While sixty-eight consultations affected projects over the seven and a half year period we examined, for the most part these effects have not been major. Further, even when the consultation affected the project, Department of the Interior and other agency officials indicated that other events occurring at the same time (such as difficulties in arranging project financing) sometimes had a more significant effect than the consultation process. The willingness and the ability of the Service and project sponsors to arrive at compromise solutions when conflicts occurred also contributed to reducing the consultation requirement's ultimate effect on project development."

Still, opponents rightly argue that the consultation process offers environmentalists another roadblock they can throw up in front of a project if they want to test the staying power of its supporters. Jim Riley, executive vice-president of the Intermountain Forest Industry Association, is the hired gun of the forest-products industry in Idaho and Montana. From his office in Coeur d'Alene, Idaho, to the offices of western congressional delegations or the Department of Agriculture in Washington, D.C., he has effectively brought home the bacon in the form of higher forest "allowable sale quantities" for timber and weakened environmental regulations. He is a professional timber beast and proud of it. His job is to clear away the roadblocks. By the end of the 1980s, the Endangered Species Act was becoming one of the major tools for slowing down the timber-cutting ardor of his employers. "It's a procedure game," Riley says. "Always extend the procedure to delay the action. It's abusing the process."

With Riley and others like him having the ear of most western congressional delegations, procedure is all that environmentalists had during the decade to throw in front of the steamroller. It wasn't environmentalists who were abusing the process, it was powerful politicians and bureaucrats working at the behest of powerful special interests, who had billions of dollars at stake, who manipulated the law. Just ask Lorraine Mintzmyer.

Mintzmyer, the former Rocky Mountain regional director of the National Park Service in Denver, says these people are not bad people. "They are not all the ecological robber barons of a generation or two past," who plundered the West with little regard for the people and land they harmed. "They may not do great wrongs, but they are part of a hidden system of many small wrongs, which add up," Mintzmyer said in a speech after her forced retirement in 1992. She was reassigned from this most coveted region to the Park Service's answer to Siberia— the eastern regional office in Philadelphia—after testifying that the White House improperly intervened in the writing of a "vision report" that was designed to improve environmental management of the Greater Yellowstone area. But that's a story for a later chapter.

142

From her Denver office, Mintzmyer was privy to many of the endangered species battles in the northern Rockies during the Reagan and Bush administrations. She had been chairwoman of the Interagency Grizzly Bear Committee that coordinated grizzly bear management in the West and was deeply involved in the controversy over reintroducing wolves to Yellowstone National Park. Mintzmyer described herself as a participant in a process that she believes threatens national parks and endangered species.

"A senator mentions that a particular matter should be addressed; a political appointee mentions that the White House wants to help some of the special interests; a regional director, superintendent, ranger or scientist changes a number, agrees to a permit or takes a section out of a report," Mintzmyer said, describing the process. Perhaps her most chilling charge, though, was that political interference made virtually every scientific document written by federal land and wildlife management agencies after 1983 suspect. Mintzmyer won't say why that date is significant, she says, because of pending litigation concerning her retirement. She said the American public couldn't be assured that the scientists' research and best judgment hadn't been manipulated or even removed from the final report of the agencies on which decisions were finally based. The manipulation of the listing process has been shown in the cases of the northern spotted owl, Bruneau Hot Springs snail and salmon. The same manipulation was operative when high-profile biological opinions were written in the 1980s.

Perhaps the highest-profile series of biological opinions in the northern Rockies have been issued concerning development around Yellowstone Lake in Yellowstone National Park. National Park Service biologists have known for nearly thirty years that development around Yellowstone Lake has forced conflicts between humans and grizzly bears. Yet in the last thirteen years along the lakeshore, the Park Service, with the approval of the FWS, has built one new development, renovated another and left the controversial Fishing Bridge development in the middle of one of the most important pieces of grizzly habitat in the park. The agency that presents itself as the most preservation-oriented in the federal government has backed off promises it made to mitigate the effects of its developments, forcing bears and visitors together at the expense of the bears. The FWS has gone along, repeatedly backing off and letting the Park Service proceed with the developments. "The Park Service has had a very difficult time

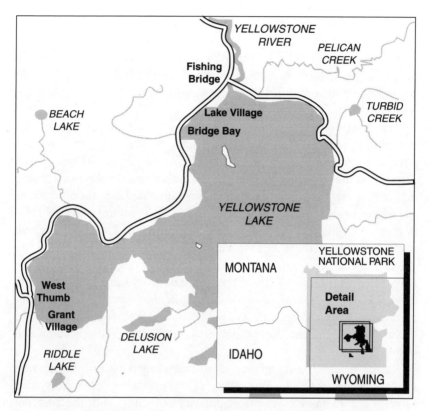

The Yellowstone Lake area, where development has placed humans in the middle of prime grizzly habitat.

acknowledging, when its data is sound, what kind of changes in recreational facilities are needed to recover the bear," says Louisa Willcox, program director of the Greater Yellowstone Coalition, based in Bozeman, Montana. "They failed to use their own data in Fishing Bridge, they failed to use the same data at Grant [Village] and now it's the same at Lake. The science is rarely so clear as it is in this situation."

The controversy began in 1979 over the proposed development of Grant Village on the south end of the lake. Grant Village was supposed to take the place of both the Fishing Bridge and Old Faithful developed areas. A master plan developed by the National Park Service in 1974 had recommended closing visitor services at Fishing Bridge and phasing out overnight accommodations at Old Faithful, the busiest spot in the park. The proposed Grant Village site also was a popular travel

area for grizzlies, especially during the cutthroat trout spawning sea-son, when bears flocked to the lake's feeder streams to fish. Before the
144   Park Service could build Grant Village in the middle of prime grizzly bear habitat, it had to formally consult with the FWS. In its 1979 biological opinion, the FWS pulled no punches about its view of the wisdom of building Grant Village.

"We question the need and justification for such extensive commer-cial development within occupied grizzly habitat, and believe that adverse impacts to the bear will result," said James Gritman, FWS acting regional director in Denver. "We also believe the project will negate many of the benefits acquired through the phase-out of facilities at Fishing Bridge, and view such a 'trade-off' as an unfavorable solu-tion to a wildlife conflict that with development of Grant Village will likely be duplicated rather than eliminated."

Despite the reservations expressed in the biological opinion, the FWS ruled that Grant Village development would not jeopardize the existence of grizzly bears if all of Fishing Bridge's services and accom-modations were closed. In 1982 the Park Service proceeded to build Grant Village. In the meantime, Fishing Bridge facilities remained open.

Fishing Bridge is a campground and developed area on the eastern side of Yellowstone National Park at the north end of Yellowstone Lake; it got its name from the popularity of fishing from the adjacent bridge that crosses the Yellowstone River. The bridge was built in 1902, and the first automobile campground was developed there in 1921. Later a cafeteria, store, museum, amphitheater, gas station, tourists cabins, employee housing and other government buildings were added. By 1979 the tourist cabins were closed because of their disrepair and 670 camping and recreational vehicle sites remained.

Not only is Fishing Bridge a popular tourist area, it is also extremely popular with the park's grizzly bears. The campground had been built smack dab in the middle of one of the most important crossroads for bears and other wildlife in Yellowstone. Bears use the area to travel between the Pelican and Hayden valleys, two important feeding areas. Bears also had historically used the area, like humans, as a major fishing spot, attractive because of the river's cutthroat trout. Heavy fishing pressure and artificial introduction of other trout species had left the cutthroat population stunted in size and numbers. The Park Service initiated a program to enhance wild stocks of fish, which included strict limits on cutthroat and closures during spawning, when the fish are most vulnerable. The program has been a major success, increasing both the size and numbers of cutthroat in the river and the lake. When cutthroat populations rebounded in the 1970s, bears increased their seasonal fishing and traveled through the area in greater numbers toward Yellowstone Lake.

Historically, Fishing Bridge had been a place where humans and bears mixed with unhappy results. Half of the forty-nine deaths to grizzly bears from 1943 to 1959 took place at Fishing Bridge and nearby Pelican Creek. It was clear to National Park scientists that Fishing Bridge was one of the most ecologically diverse and important areas in Yellowstone. It also was clear that human development didn't belong there.

In the 1974 master plan, an environmental impact statement recommended eliminating "accommodations and services from this existing developed area in order to facilitate restoration of critical wildlife habitats at Yellowstone Lake's outlet." It went on to say: "The area from the mouth of the Yellowstone River to one mile downstream is superb ecological environment and should be restored to its natural condition."

James Watt, secretary of the interior from 1981 to 1983, said that when Grant Village was approved, Yellowstone's superintendent, John Townsley, came to him and outlined his plan to build Grant Village and close Fishing Bridge. Watt, a fierce defender of the free market, was uneasy about a government-sponsored development that would compete against private businesses outside the park but gave his go-ahead. He told Townsley to close Fishing Bridge, but then Townsley and the Park Service failed to move decisively on the issue. Its delay, Watt said, allowed the Wyoming congressional delegation time to enter the debate. The delegation was led by Senator Alan Simpson, a former Cody, Wyoming, lawyer who had heard from his hometown chamber of commerce that closing Fishing Bridge would be devastating to the tourism economy of the gateway community on Yellowstone's eastern boundary. Environmentalists said Cody's opposition to closing Fishing Bridge was puzzling, since people who couldn't find accommodations in the park might be lured to stay in Cody. Even though Grant Village was built to take up the slack for Fishing Bridge, and despite the Park Service's commitment to replacing the camping facilities elsewhere in the park, Simpson wanted the closure of Fishing Bridge reconsidered.

146

He made a formal request in 1984, after Watt had left his position under fire. Watt no longer could provide political cover for Townsley. The Department of the Interior was in a mild state of drift following the very ideological leadership of Watt. So with Reagan advisor William Clark in control, it reverted to a more partisan basis for policy and put pressure on the Park Service to meet Simpson's demands.

In July 1985, regional director Mintzmyer informed the Fish and Wildlife Service that the Park Service was going to honor the request of the Wyoming delegation and conduct a full environmental impact review of the closing of Fishing Bridge and would not honor its agreement to close Fishing Bridge by the end of 1985. (Perhaps when Mintzmyer referred to the "hidden system" in 1992, she was remembering this major concession to Simpson and the powerful special interests he represented.) She requested reinitiation of formal consultation and said that a decision would be reached by 1987.

An interim management plan was proposed that left both Grant Village and Fishing Bridge open with additional constraints on day use and other measures to reduce the conflicts between humans and bears. On May 7, 1986, the FWS issued its biological opinion on the interim plan, an opinion that was a complete reversal of the 1979 Grant Village opinion. The new opinion said that simultaneous operation of Fishing Bridge and Grant Village would not jeopardize the existence of the grizzly, and that decision would be in effect only until Yellowstone finished its Fishing Bridge environmental impact statement.

In October 1985, FWS regional director Galen Buterbaugh had told the Park Service to prepare a biological assessment before formal consultations resumed. He suggested analyzing the cumulative effects of development using a new, untested computer model designed to measure the effects of activities on grizzly habitat.

The Park Service decided to close the 310-site campground and several of the commercial operations at Fishing Bridge. But the 360-site recreational vehicle campground would stay, along with all historic structures. Additional measures to protect grizzlies, such as seasonal trail closures, would be implemented. The computer model indicated that all of the proposed actions would result in a net increase of 1500 acres of prime grizzly habitat. Later, however, the cumulative effects computer model was universally criticized by grizzly bear biologists as inaccurate and inadequate. In their biological opinion of October 13, 1987, FWS biologists estimated that the proposed actions at

Fishing Bridge would save an additional 1.1 bears in the following decade. That would make up for the .07 bear expected to be lost at Grant Village, said the FWS. So Fishing Bridge could remain, Grant Village could remain and the Park Service could even build another campground if it chose. The FWS had backed off from its strong position to protect the bear in 1979.

Buterbaugh, who, like Mintzmyer, was transferred from his Rocky Mountain regional director job, says he knew of no political pressure on his scientists, who were consulting with the Park Service. "The Park Service was getting all the pressure," Buterbaugh says. "I knew they were getting it, but we weren't."

Wayne Brewster, the main biologist on the Fishing Bridge project for the FWS, confirmed Buterbaugh's assessment. Yet his own description of events showed that the Park Service was steering the process by offering a continuous stream of alternatives, to which his scientists would respond. Brewster said this is the proper method of conducting biological opinions, allowing the agency proposing an action to seek alternatives to harming species.

Yet when the process is driven by politics, as the Fishing Bridge controversy was, stamina becomes an important factor. Facing political consequences, as well as a dichotomous mandate between preservation and tourism promotion, the Park Service was unwilling to give any space to the bear it didn't have to. The FWS was charged with telling the Park Service how much space it had to relinquish. After a while in any negotiations it gets harder to say no, especially when dealing from the position of weakness accorded biologists working for the Reagan and Bush administrations' Department of the Interior. No one questions that grizzly bears would be better off had Grant Village been left unbuilt and Fishing Bridge closed. But where was the proper balance between bears and humans?

Yellowstone superintendent Bob Barbee has to balance, on a daily basis, the need to protect the park's resources and the people's right to view them. The large, jovial Barbee cuts a striking figure when decked out in full ranger uniform. He takes great pride in the Park Service and pays a great deal of attention to tradition, history and honor for the preservationist past on which the Park Service is built. Barbee is one of the Park Service's top people. He is rumored to have turned down a regional director post so that he could stay at Yellowstone, the flagship of the Park Service. He also is one of the most politically savvy natural

148

resource managers in all of the federal agencies. When Yellowstone burned in 1988, many western senators and representatives characterized Barbee as a modern-day Nero, fiddling as hundreds of thousands of acres burned. Several called on the Park Service to fire him. Barbee not only survived, but successfully defended the Park Service's natural fire policy when it was reviewed.

Barbee also has displayed political courage, standing up in defense of Yellowstone's geothermal features in 1991 when the Church Universal and Triumphant wanted to pump water from a hot spring only a couple miles north of the park boundary. Interior secretary Manuel Lujan had approved the well, based on a U.S. Geological Survey report that said there was little chance it would hurt famous park geothermal features, such as Mammoth Hot Springs. Barbee challenged that report and in a letter told Lujan that if they were to err, they should err on the side of preservation.

Yet on Fishing Bridge, a matter that he inherited from Townsley when he arrived in 1984, and later on, in the associated Lake–Bridge Bay decision in 1992, Barbee appears to have leaned toward development.

The historic Lake Hotel is located in a developed area only a couple of miles southwest of Fishing Bridge on the shores of Yellowstone Lake. The area also has tourist cabins, a small hospital, a smaller lodge, several stores, a ranger station and an old fish hatchery. Bridge Bay is another five miles south of Lake and includes a marina, ranger station and campground. In keeping with the bureaucratic bait and switch that allowed Grant Village to be built and Fishing Bridge to stay open, Lake and Bridge Bay were designated as the places to move facilities removed from Fishing Bridge. But before the Park Service could do anything it had to write up a plan, conduct an environmental impact review and determine the plan's impact on grizzly bears.

The Lake–Bridge Bay Development Concept Plan calls for moving an automobile service station and employee housing from Fishing Bridge to the Lake area, building a new motel facility at Lake Lodge to replace cabins along Lodge Creek, construction of a fire station and renovation of other buildings. It also calls for rehabilitation of former grizzly habitat and renovation of the Bridge Bay Campground.

In June 1992, the FWS took issue with the plan. The Park Service said the proposal would not adversely affect the grizzly bear, a threatened species protected by the federal Endangered Species Act. Under the law, the FWS must concur with the Park Service before the plan can move forward. Charles Davis, an FWS biologist in Cheyenne, wrote that the agency did not agree. Davis raised the issue of seasonal opening dates in a letter on June 1. "We are concerned with the continuing juxtapositioning of significant human activity on top of a bear use area, especially in view of the bear management actions and removals that have resulted in the past," Davis wrote. At issue were two creeks that generally act as boundaries for the Lake area. Lodge and Hatchery creeks both fill with spawning cutthroat trout in the spring, attracting grizzly bears. In 1990, Park Service officials moved a female grizzly to a research center because she regularly fed at Lodge Creek despite the presence of humans.

Daniel Reinhart, a Yellowstone biologist, and David Mattson, a biologist with the Interagency Grizzly Bear Research Team, wrote in a paper released in 1990 that human activity within one kilometer affects bear use of cutthroat spawning streams as a food source. "These effects could be mitigated by reducing the temporary overlap of spawning runs and human use of developments," Mattson and Reinhart wrote. "Given that bears are better able to adapt to predictable human behavior, a fixed but late opening schedule for lakeshore developments may be required for effective mitigation."

After the initial disagreements, Davis and other FWS biologists met with park officials and ironed out their differences. Davis said the opinion set guidelines for opening the area around Lodge Creek but did not require a delayed opening of the Lake Hotel. That made environmentalists unhappy. "The Park Service's own scientists tell us the only way to resolve the problem is to push back the date of the opening," says Louisa Willcox, Greater Yellowstone Coalition program director. "Our intention is to use the best science, the only science for a sounder program that doesn't jeopardize tourists and doesn't jeopardize bears."

Barbee says he listened to the opinions of his scientists and the coalition, but says he has a larger responsibility than they do. "We're trying to balance public use and resource preservation," he says. "They're only interested in preservation." Barbee says that the Lake developed area is "a place for people," not bears, and he has no intention of changing that. "Trying to make the Lake developed area a place for people and bears leads to conflict," Barbee says. "We believe we can manage the bear use of the area when it begins to impinge on the dedicated use of that area, which is essentially for people."

150   Environmentalists filed a notice that they intend to sue over the Lake–Bridge Bay decision, drawing a line in the sand on the Endangered Species Act and the bears of Yellowstone Lake. If the case ultimately ends up in court it could take years to resolve. So far, the issue has not jumped into the political arena, at least not publicly. Yet after years of political meddling in the Yellowstone Lake bear controversy, perhaps it wasn't necessary to make the gun-shy agencies lean toward development. The Park Service and the FWS no longer are challenging the popular view of surrounding communities and state congressional delegations that development must continue despite its effects on bears. The political pressure in the Fishing Bridge case was overt, but on the face of it conducted through the proper channels. What's missing is a paper trail that takes the decision down to ground level. Up till now, no one has linked the Wyoming delegation's request for an environmental impact statement with any specific doctoring of the facts to meet the outcome.

But another consultation involving grizzly bears in northern Montana clearly carries political footprints down to the final decision. The issue was a proposed timber harvest program in the Yaak River drainage in the extreme northwest corner of Montana. The Forest Service wanted to cut 151 million board feet of timber from 9821 acres of forestland. To do it would have required fifty miles of new road and the reconstruction of another thirty-three miles.

The problem was that the area is one of the most important pieces of grizzly bear habitat in Montana. It is part of the 2600-square-mile Cabinet-Yaak Ecosystem, which has been designated a recovery zone for grizzlies. This key mountainous, forested area lies on the Montana-Idaho border and reaches north into British Columbia. To the west lie the Selkirk Mountains, another smaller bear recovery area. To the east is the larger Northern Continental Divide Ecosystem recovery area, which includes Glacier Park and the Bob Marshall Wilderness. The Cabinet-Yaak Ecosystem provides a key linkage zone between the other ecosystems as well as being important grizzly habitat on its own.

The Yaak area also is an important timber-producing area and has been for years. Sawmills in communities such as Troy, Montana, have depended on its productive pine, fir, spruce and hemlock stands for wood supplies. Hundreds of jobs in isolated communities are tied directly to these timber harvests. Most of these communities have few of the alternatives that their neighbors nearer to parks and recreation centers have for economic development.

No one knows how many grizzly bears live in the Yaak River area, but FWS biologists determined that the numbers of bears had been declining during the 1980s. And it was no wonder. All but one of the six management units in the Yaak area were determined to be below minimum standards for grizzly habitat due to excessive roads and clearcuts. In other words, the area was so heavily clearcut and criss-crossed with roads that bears could not or would not live there. The Forest Service had ignored its own rules for leaving strips of timber between clearcuts, so that large openings averaging 156 acres apiece cover much of the area. More than 183 miles of streamside areas, the most important habitat for bears, were harvested.

"Due to the above past timber harvesting practices, security for grizzlies has been greatly reduced in the Yaak Valley, making bears vulnerable to illegal killing and susceptible to displacement from human activities," John Spinks, FWS deputy regional director wrote in a May 3, 1990, biological opinion stating that the Forest Service's recently submitted plan would jeopardize the existence of the grizzly bear. In that opinion, the FWS said the current conditions in the Yaak jeopardized the bear and the only way more timber could be harvested was through an aggressive, immediate road-closure program to make up for past logging effects.

Despite this jeopardy opinion, the FWS and the Forest Service were able to reach an agreement on an alternative that dramatically lowered the number of acres affected and included immediate road closures to mitigate the effects on bears. The agreement was based on the fact that the Forest Services's original Yaak timber-cutting plan, and all of the alternatives it presented, didn't even meet the Kootenai National Forest standards, outlined in its land management plan. The FWS and the Forest Service decided that the way to resolve their differences was to bring the Forest Service in compliance with its own rules for protecting grizzly bears. Only the alternative known as "9b," which permitted the harvest of 65 million board feet, met these rules.

"It was agreed and desired to meet all standards and intents before any activity is approved," Spinks wrote. "The result, alternative 9b, has been developed to meet all previous concerns with all publics and agencies, and to fully accommodate these later items."

152

At least two of the agencies' "publics" weren't satisfied with the agreement: the timber industry, represented by Jim Riley, and Senator James McClure (R-Idaho). Their opposition was not based on simply trying to squeeze more wood off the Kootenai. Both the industry and McClure believed philosophically that they had to act immediately to avert the waste of millions of board feet of marketable timber. The western pine beetle had infested the lodgepole pine forest of the Kootenai the same way it had infested millions of acres of lodgepole throughout the West. McClure and the timber industry wanted to cut the standing lodgepole immediately, before it was destroyed by the beetle. Following the scientific forestry philosophy, the area would be reseeded, and a new, more efficient forest would replace the sick, inefficient one. McClure believed the Kootenai staff had acted too late already. He arranged a meeting of his staff and forest-industry representatives with the Forest Service and the FWS on May 24, 1990. McClure wasn't happy with the interagency cooperation that had been reflected in the May 3 biological opinion.

In a letter to the secretary of agriculture, Clayton Yuetter, on June 12, 1990, McClure gave his version of the meeting. "In summary, the Forest Service admitted maybe they had misunderstood what the FWS had intended and the FWS admitted maybe there were other options they should have examined," McClure wrote. In other words, McClure was saying, this appeared to be just a little misunderstanding that, with a little help, could be worked out his way. He told Yuetter that the Forest Service should have simply presented what it considered the best management plan, let the FWS issue a jeopardy opinion and, instead of negotiating, forced the FWS to rewrite the harvest plan.

"He [McClure] was setting up a shootout at the OK Corral," says Keith Hammer, chairman of the Swan View Coalition, a local environmental group that opposed the timber harvest. "McClure called for interagency confrontation."

McClure said the restrictions on timber harvest weren't needed to protect the grizzly bear and that the real issue was the Forest Service's unwillingness to fight for more timber cutting.

"The Forest Service is tired of the battle and finds it easier to roll over than to do a professional job in the face of stiff and unrelenting opposition to the use of public lands for commodity production," McClure wrote Yuetter. "This administration can do something about that and I hope you will. I will certainly help in any way I can."

McClure had done enough. The FWS decided it would not require the roads to be immediately closed, and it loosened several other constraints that increased the harvest to 90 million board feet in alternative "9a." Because of the road-closure delays, the new plan wouldn't meet the requirements of the law until after the timber had been cut. The two agencies' new plan was unveiled in a biological opinion issued June 20, 1990, eight days after McClure sent the Yuetter letter and less than a month after his meeting with Forest Service and FWS personnel.

"In choosing to increase logging levels the agencies knowingly decided not to meet a number of essential grizzly bear protection standards during the life of the five-year timber program," Hammer says. "This violates both prior agreements and prior biological opinions. The Fish and Wildlife Service allowed the consultation process to turn into a penny-ante game where political pressure is trump over science and professional integrity."

Section 7 of the Endangered Species Act, besides requiring that federal actions don't jeopardize listed species, also requires the designation of critical habitat for species—the land or water that must be protected for their survival. Critical habitat can include private land and even water owned under state law by farmers and other users. In essence, destroying critical habitat—even on private land—would be considered the same as destroying the species, and violators would face stiff fines and jail. Because of this particularly harsh measure, the Endangered Species Act specifically allows economics to be considered in the designation of critical habitat.

The FWS has failed to designate critical habitat for hundreds of species, including such high-profile animals as grizzly bears. Of the 651 species listed in May 1992, only 105—16 percent—had critical habitat designated.

In the case of the grizzly bear, in 1976, the FWS actually proposed designating 20,000 square miles, mostly public land, in Wyoming, Idaho and Montana as critical habitat. But westerners overflowed hearing rooms in their efforts to express opposition to the proposal they saw as a threat to their way of life. Many scientists agreed with the opponents at the time, arguing that not enough was known about where grizzlies lived, nor how many there were and how much space they needed.

"Habitat designation is premature at this time," said Chris Servheen in a hearing in Missoula, Montana, in 1976. Servheen was a research biologist working on a joint U.S. and Canadian grizzly bear project. Today he is the FWS's grizzly bear recovery coordinator, and he remains stingy when it comes to adding protected habitat to the recovery areas for bears. When grizzly-occupied habitat was identified in the early 1980s and the bear recovery plan written, biologists dodged the issue of grizzly habitat designation again, so today activities on private land continue with little or no attention to their effects on bears. Perhaps the most dramatic result of this historic reluctance to designate grizzly habitat is the loss of thousands of acres of prime habitat in the 1980s clearcut by the Plum Creek Timber Company, the largest private owner of grizzly habitat in the United States. Designation of critical habitat would have forced Plum Creek to cut timber in a manner that protected bears instead of the section-size clearcuts that have fragmented much of Montana's best bear country.

In 1989, the FWS also wanted to avoid designating critical habitat for the northern spotted owl. But environmentalists returned to Judge Thomas Zilly's U.S. District Court in August 1990 to ask that he order such designation. He did in February 1991, and in January 1992, the agency designated 6.9 million acres of federal land in Washington, Oregon and California as critical habitat. To dodge the heaviest criticism from the timber industry and woodworkers, the FWS deftly left out 4.7 million acres of private and state lands identified earlier as important owl habitat.

The consulting process works differently for species that are nearer to extinction than it does for those nearer to recovery. When a species is listed as threatened, some loss of individual animals or plants is allowed. But when a species is endangered and close to extinction, each individual organism is considered important to the future of the species.

When the northern spotted owl was listed in 1990, the FWS faced a formidable consulting task. It had to review thousands of timber sales in Oregon, Washington and California on federal, state and private lands to determine their effects on the owl's habitat. "I have biologists that have put in one hell of a year, hugged their wives, kissed kids goodbye and left for weeks and months at a time, worked seven-day weeks and fourteen-hour days to try and meet their workload," said John Turner, former FWS director. "We've turned those sales around in thirty days, much quicker than the law allows, and that's a lot of long nights. The job they have done is incredible."

The owl consulting program kept FWS biologists busy, mostly with the Forest Service. Meanwhile the Bureau of Land Management (BLM) released its own two-year timber sale plan and decided to ignore the requirement to consult with the FWS. Environmentalists took the BLM to court in April 1991. In September of that year, U.S. District Court Judge Robert Jones ruled that the BLM had violated the Endangered Species Act by failing to consult.

The FWS had issued a jeopardy opinion on forty-four timber sales on Oregon BLM lands. BLM director Cy Jamison, frustrated with the power now in the hands of the FWS over his timber program, chose to test the little-used exemption process to override the powers of the act.

The exemption process was added to the Endangered Species Act as a result of the Tellico Dam decision involving the now-famous snail darter. The process is designed to allow economics to be taken into account along with the survival of a species. Only through the exemption process can the costs of protecting a species overrule its protection. But the process is complicated, and the test a developer or federal official must overcome is hard.

Under the exemption process, an Endangered Species Committee is composed of the secretary of agriculture, the secretary of the army, the chairman of the Council of Economic Advisors, the Environmental Protection Agency administrator, the National Oceanic and Atmospheric Administration administrator and one individual from each affected state.

The problem for the Pacific Northwest timber industry is that the procedure can be triggered only by a jeopardy opinion on a specific project, such as one timber sale. Senator Robert Packwood (R-Oregon) introduced a bill in the Senate in 1990 that would have allowed the committee to balance the economic effects on a regional basis instead. The proposal was defeated in the Senate and had even stronger opposition in the House.

This requirement makes the use of the exemption process for skirting the northern spotted owl issue even more difficult, as Jamison found out. The Endangered Species Committee, almost universally dubbed the "God Squad" for its sweeping powers to allow a species to go extinct, must meet four tests before it rules against a species: (1) there are no reasonable and prudent alternatives to an agency's proposed species-threatening action; (2) the benefits of such action clearly outweigh the benefits of alternative courses of action consistent with conserving the species or its critical habitat, and that such action is in the public interest; (3) the action is of regional or national significance; (4) the agency applying for an exemption had not already made an irreversible commitment of resources toward the species-threatening action so it could hold the species hostage to these already-spent costs.

Ruling on the national or regional significance of a timber sale, or even forty-four timber sales, would be hard. And the test of reasonable and prudent alternatives remains as hard in the exemption process as it is in the consultation process.

In May 1992, the God Squad issued an ambiguous decision that demonstrated the weakness of the exemption process. It voted 5–2 to exempt thirteen of the forty-four sales from the Endangered Species Act—but not until the BLM developed a scientifically sound plan for protecting the spotted owl.

To add insult to injury, the thirteen sales remained tied up in court battles over the BLM's lack of compliance with other federal environmental laws. In the end, if the sales go through, they will keep 1000 timber workers on the job for an extra year. It simply puts off the bust for another year in desperate rural communities in southern Oregon.

Where the exemption process holds more promise, or provocation, depending on your point of view, is in the salmon controversy. The regional nature of the controversy, covering 200,000 square miles of watershed, makes salmon fit the exemption process like a glove.

The most significant effects will be on the complex Federal Columbia River Power System and other river users connected to its management, including irrigators, barge operators, the Bonneville Power Administration, Army Corps of Engineers, Bureau of Reclamation and the Idaho Power Company, which operates its own hydroelectric dams on the Snake River.

When the National Marine Fisheries Service initiated formal consultation on the power system, all of these users were taken into account. The private entities—irrigators, barge operators and Idaho Power—are involved through federal agencies that control their water and operations. Irrigators, for instance, store water in Bureau of Reclamation reservoirs in Idaho and Wyoming. Idaho Power operates its dam under a Federal Energy Regulatory Commission license.

Prior to the listing of sockeye salmon in 1990, the NMFS Northwest regional staff had no need to consult, since it had no threatened or endangered species. Nationwide, the agency has not had anywhere near the consultation experience that the FWS has. From 1987 through 1991, the NMFS had 691 informal consultations and 244 formal consultations. In that time, only three jeopardy opinions were issued. 157

In the 1990s, the NMFS faces the most daunting consultation task in the history of the Endangered Species Act. In 1992, it had six employees to carry out what would be thousands of individual formal and informal consultations on everything from timber sales, mining permits, marina projects, water projects and commercial fishing harvests to dam operations. The salmon listing "created a real avalanche of work," says Merritt Tuttle, who heads the effort for the NMFS. "You've got a whole mass of items coming in daily and most take months to go through."

Even if the consultations were routine, it would be an unprecedented workload. But many of the consultations will be extremely controversial, and therefore regularly attracting political interference. The NMFS's first major biological opinion was perhaps a harbinger of the battles to come in the salmon controversy.

The NMFS had decided that it could not immediately provide one all-inclusive biological opinion for the Federal Columbia River Power System. Bruce Fox, former NMFS director, said that eventually, when a recovery plan is completed, it would do one comprehensive consultation, but in the meantime, it would issue biological opinions on an annual basis for the river management system.

In its first biological opinion, the NMFS shocked environmentalists by issuing a no-jeopardy opinion. "We conclude that the proposed operations are not likely to jeopardize the continued existence of listed or proposed salmon species," Fox wrote on April 10, 1992. "Nonetheless, we are concerned that if operation of FCRPS continued as it is proposed for 1992, it would not be sufficient to reverse the decline over one life cycle of the salmon; therefore, additional steps will likely be needed in 1993 and future years."

Fox issued the no-jeopardy opinion even though the BPA and the Army Corps of Engineers acknowledged that their dam operations would kill up to 82 percent of Snake River juvenile salmon migrating to the ocean. Sixty-six percent of returning adult salmon also would be allowed to die under the plan.

158

Like Yellowstone's Grant Village biological opinion, the NMFS salmon opinion allows the agency to criticize its fellow federal agencies and outline what they were doing wrong without stopping them from doing it. It delays the day of reckoning and gives hope to those who think that simply tweaking the system around the edges might be enough to get them around the Endangered Species Act with regard to salmon.

However, the no-jeopardy opinion sparked lawsuits from all quarters. Environmentalists sued because the opinion allowed the hydroelectric dams to continue killing fish. Utilities, irrigators and industrial river users sued because the NMFS plan wasn't tough enough on tribal fishermen, upstream logging operations and other salmon killers.

In that atmosphere, a federal agency such as the Army Corps of Engineers might find it attractive eventually to take the whole issue to the God Squad. But like the timber industry and the BLM before it, the Corps or other impatient land and river managers could well be disappointed by the outcome, since many other environmental laws, especially the Northwest Power Act, could override the Endangered Species Act exemption.

The bottom line is that there are "reasonable and prudent" alternatives for saving salmon, just as there are for most endangered species. Despite this, political pressures will always steer the agencies toward the easiest route to protecting species and their habitats. As the examples of Fishing Bridge, Yaak Valley and Snake River salmon demonstrate, the language of Section 7 may be airtight, but the actual enforcement is flexible to the point of breakage.

7. Roads to Recovery

BILL HEINRICH dangled by a rope more than a thousand feet above the floor of Palisades Creek Canyon in the Snake River Mountains on the Idaho-Wyoming border. In his hands he held a piece of the future of the peregrine falcon—five young falcons bred in captivity.

Heinrich, a soft-spoken raptor biologist for the Peregrine Fund, was restoring—"hacking," as Heinrich called it—the feisty fledgling peregrines to the wild in an artificial nest site—a desk-size hack box—that would be their home until they got their wings. Heinrich is one of a small corps of biologists and volunteers who have created one of the great success stories of the Endangered Species Act. The Boise-based Peregrine Fund has returned healthy populations of peregrine falcons to the wild in the United States, where they had been nearly exterminated by agricultural chemicals.

Removing a species from the Endangered Species List can be the mark of success—or resounding failure. Since the Endangered Species Act was passed in 1973, sixteen species have been delisted, officially removed from the list. In four cases, the original data used to warrant listing were in error; in seven other cases, the species were determined to be extinct. But five have been official success stories, and a sixth, the American alligator, is recognized as recovered and remains on the list only because it looks like other listed species.

The peregrine falcon hasn't been delisted yet, but it could well be before the end of the decade. Its success story is a stark contrast to most endangered species stories of the last twenty years. And the recovery has been relatively easy. Since habitat destruction did not lead to the disappearance of peregrine falcons, habitat protection is not a serious issue. There were millions of acres of vacant habitat waiting to be filled. Protecting peregrines didn't require the lock-up of vast areas of habitat. The bird's survival would not stop anyone from building anything. Peregrines could even be turned into city-dwellers, living on the ledges of skyscrapers and feasting on the ever-abundant pigeons. Peregrines were an endangered species that any corporate CEO could love.

Most endangered species became endangered because their lifestyle doesn't fit in with modern men and women. Grizzlies need a lot of space and are downright ornery when they don't get it. Wolves can't keep their teeth out of a tasty steer if available, and salmon need clean water, a clear path home and at least a few holes in the worldwide fishing net to survive. All falcons need is a good ledge for a nest, a good prey base and a poison-free environment. It was this third requirement that almost led to extinction for the peregrine falcon in the late 1960s.

Peregrines, like their raptor cousin the bald eagle, were seriously depleted by widespread pesticide use following World War II. DDT attacks the reproductive system of raptors and weakens the shells of their eggs. Falcons, like eagles, are at the top of the food chain. By the 1960s, DDT concentrations in falcons and eagles were especially high, since they had accumulated the concentrations of their prey, which had accumulated the concentrations of their food. In essence, falcons and other raptors were an early warning system for the entire world biosphere. Continued use of DDT would eventually have destroyed the web on which all life depends. In 1972, it was banned in the United States.

When DDT was banned, scientists weren't sure that the government had acted soon enough for peregrines. After 1968 there were no wild falcons left east of the Mississippi. In the West, 80 to 90 percent of the population was gone. A viable population remained in Alaska and northern Canada, but everywhere else the screeching cry of the falcon no longer could be heard.

In 1970, Dr. Tom Cade, an ornithologist at Cornell University, began experiments designed to propagate captive falcons. To support his program, Cade established the Peregrine Fund. By 1973, when the Endangered Species Act was passed, Cade had raised twenty falcons. In 1974, Cade's team was experimentally returning falcons to the wild.

The first successful hacking took place in 1976 in New York's Hudson Palisades area.

Peregrines are a crow-size raptor with a sleek, compact body and long, pointed wings. Peregrines can cruise at 40 to 60 miles per hour or plunge toward the ground at 200 miles per hour or more, making them the fastest of all raptors. They feed on birds in a hunting territory that can extend ten to twenty miles from their nest. The male peregrine is the one that establishes its breeding territory, in February; the female chooses the nesting ledge. Eggs are laid in early April and hatch about a month later. The young birds begin flying in August. In the winter, peregrines fly south to Mexico and Central America.

While restoring falcons was easier than getting ranchers to accept the presence of grizzly bears it wasn't a stroll through the park. Cade and other Peregrine Fund researchers had to develop an entirely new artificial reproduction program. They had to go through a series of trial-and-error experiments both in the laboratory and in the wild before finally learning how to hack falcons. Human-raised peregrines didn't begin naturally reproducing in the wild until 1980. Henrich's experience on the Palisades Creek cliffs showed why.

Hacking is a complicated, delicate process. Nature can be a harsh nursemaid, even when humans step in to help. Young falcons, raised with tender loving care in the Peregrine Fund's World Center for Birds of Prey in Boise, must be left on the side of a cliff to fend for themselves. The stern-faced, shaggy-feathered youngsters did get help. In addition to Henrich, a pair of volunteers, Barbara Boileau of Logan, Utah, and Ann Lettenberger of Hood River, Oregon, climbed out on the ledge towering above Palisades Creek to help the birds set up housekeeping. Because hack sites usually are lonely cliffs reached only by technical climbing, the hacking team had to be skilled in rock climbing. They also had to be sensitive foster parents, feeding the fledgling falcons for up to eight weeks before the birds learned to hunt on their own. It was lonely, often monotonous duty for the volunteers Henrich left behind. The baby-sitters kept track of the peregrines with receivers that picked up radio signals from tiny transmitters placed on the birds. Each day they would climb up to the hack site and leave food for the birds, carefully staying out of sight so that the birds didn't become accustomed to human disturbance.

When everything goes right, the birds begin flying after about a week in the hack box. Then they become easy prey for golden eagles and

161

great horned owls. The foster parent volunteers can help ward off those predators but it is hard. In addition, the young birds are suscept-

162   ible to viruses and other diseases.

Within ten days of release, four of the five fledglings—80 percent— had died of disease or predation. The remaining bird was returned to Boise. "We're always shooting for 100 percent, but nature doesn't work that way," Heinrich said afterward, disappointed but undaunted.

Henrich's attitude is justified. The research and attention to detail has made the Peregrine Fund extremely successful at hacking. Since the Peregrine Fund begun reintroducing falcons in 1974, more than 3400 have been released in twenty-seven states. Of those, 82 percent were known to have survived. In 1991, at least 700 pairs of peregrines had set up territories in the Lower 48 states.

The Peregrine Fund's reintroduction program east of the Mississippi was ended in 1991, with victory declared. The group recommended that the U.S. Fish and Wildlife Service (FWS) begin a timetable that would lead to downlisting of peregrine falcons to threatened status in 1995 and delisting by the turn of the century. In the West, the group recommended immediate delisting in all states but Idaho, Wyoming, Montana, Washington and Oregon. Even in those states, the fund's biologists recommend that delisting could be justified as early as 1996.

Despite the strong evidence of recovery, the FWS is not racing toward delisting. In fact, the agency is dragging its feet. "The pressure is to maintain the program at the current level," says Rich Howard, a raptor biologist for the agency in Boise, Idaho. Part of the reason is that almost everyone welcomes peregrines in their backyards. From ranch lands to skyscrapers, peregrines have made many new friends for themselves and for endangered species in general. The endangered status of falcons has meant the continued reintroduction of falcons by the Peregrine Fund. If the falcon were to be delisted, the reintroduction program, already ended in the East, would end in the West. Howard says there is a more pragmatic reason why both the federal government and the states are hesitant to push for early delisting: they don't want to blow it. Once a species is delisted, the law requires continued federal monitoring of its health for five years. But in the case of peregrines, Howard says officials want to ensure the species' long-term health before delisting.

"You do not create a recovery program as extensive as the peregrine program overnight," he says. "If we started to delist the population and break down the organization and then found out four or five years later that we have some real problems, it would be hard to recreate that program."

163

But Howard acknowledges that most of the issues that need to be resolved now are management issues, not recovery issues. These issues include ensuring that wintering sites in Mexico and Central America remain available, a growing problem since development is dramatically changing the landscape of Mexico. New pesticides also could quickly threaten the entire population if not studied carefully. "Those are management challenges, not threats," he said. The FWS is amending its falcon recovery plan to bring together two separate plans under one document, a process that Howard says is necessary to make delisting easy. But those waiting for delisting shouldn't hold their breath. Unlike the case of the grizzly bear, the states welcome the recovery efforts for falcon. The states have none of the headaches of angry landowners and public land restrictions, and all of the benefits, such as added support for nongame programs and a convenient success story they can roll out for show when they want to prove they're doing good work.

Bald eagles also have made a strong and steady comeback since DDT was banned. In contrast to the FWS's go-slow approach on falcons, the agency already has begun a process for downlisting bald eagles from endangered to threatened status, thus allowing managers more latitude when dealing with the bird and, most significantly, its more troublesome habitat needs. The difference in approach is small, but the contrast between bald eagle recovery and peregrine recovery demonstrates how pressures of habitat protection drive the delisting process. The bald eagle is the national symbol of the United States and therefore has a much larger constituency of support for its protection; yet its more burdensome management needs place pressures on managers to look for easier management options, such as giving up a nesting site. I his is a problem not often faced with peregrines.

Soon after bald eagles became the official bird on the emblem of the United States in 1782, their numbers began to decline. The spread of civilization westward destroyed eagle-nesting habitat, and hunters, trappers, collectors and even bounty hunters decimated the population. Still, as late as a century ago, the population was estimated at a quarter of a million. By the 1960s, only 400 pairs were left nationwide.

Today the bald eagle is listed as an endangered species in forty-three states, including Idaho, and as a threatened species in five other states. Most eagles mate for life and return to the same tree or one of several trees in the same area. They don't adapt well to disturbance in the same manner that peregrines appear to do. They nest and winter usually near large bodies of water, since fish are their favorite food. Humans also tend to like to be near large bodies of water themselves, to boat, camp, fish and build second-home subdivisions among other things. In the winter, eagles move south in search of open waters and roost together in groves of trees that shelter them from storms and other disturbances. To protect eagles, managers have to protect both their nesting areas and wintering areas. Studies show that if human disturbance regularly takes place within about a quarter of a mile of a nest, adult eagles become less attentive to their young or even will leave, leading to the death of at least some of the young birds. Human disturbance also places stress on wintering eagles and can lead to mortalities, especially in immature birds. So biologists and managers working to restore eagle numbers have to keep enough space between eagles and humans to prevent the eagles from leaving.

The first, and perhaps biggest, step in bald eagle recovery happened even before the Endangered Species Act became law, with the banning of the pesticide DDT in 1972. After that, the eagle's breeding rate rebounded significantly.

This wasn't enough for the eagles, however. There was the matter of habitat protection. Luckily, the kind of protection that the act provides best—forcing minor changes in development plans—is ideal for eagles in most cases. A timber sale can be designed to avoid a nesting area. A road usually can be moved. A campsite can be rearranged. Eagles rarely forced humans to stop their development plans.

Small adjustments in land management and the banning of DDT resulted in a phenomenal recovery. From FWS estimates of as few as 400 nesting eagle pairs in the early 1960s, the population has skyrocketed to at least 2660 pairs by 1990—greater than a sixfold increase. In the Pacific Northwest, the eagles rebounded in the 1980s. In 1980, some 276 bald eagle nests were occupied in seven states. By 1990, nesting pairs had risen to 861—more than a threefold increase.

In February 1990, the FWS formally began a process to study reclassification of eagles from endangered to threatened. The downlisting would not mean much in management terms, but theoretically managers would be allowed a little more flexibility when facing major modifications to development activities. More important to John Turner, former FWS director and a former eagle biologist himself, it would demonstrate that the act was in fact working.

"The dramatic growth of eagle populations in recent years leads us to think the species may no longer be in danger of becoming extinct," stated Turner in a 1990 press release. "It is possible a reclassification to threatened may reflect more accurately the species' actual biological status. I want to emphasize the Service is not considering removing the bald eagle from the protection of the Endangered Species Act. A re-classification to threatened, should we decide that is warranted, would continue to offer the full protection of the Endangered Species Act."

What Turner didn't say is that as eagle numbers had been rising, protection had been eroding. Karen Steenhof, a Boise raptor biologist with the Bureau of Land Management and leader of the Pacific Bald Eagle Recovery Team, issued a report in 1990 supporting downlisting of bald eagles in Oregon, Washington, Idaho, Montana, Wyoming, Nevada and California. Despite her overall optimism, she warned that the FWS was letting too many eagles die. "Agency officials are becoming dangerously complacent about the loss of individual nests," Steenhof wrote. "U.S. Fish and Wildlife Service officials are now reluctant to issue 'jeopardy' decisions, and are more apt to authorize an incidental 'take' [killing or habitat disturbance] when one or more eagle nests is threatened."

When the Army Corps of Engineers was improving a levee on the Snake River to prevent flooding near Jackson, Wyoming, the FWS allowed the agency to disturb, perhaps permanently, several of the most productive nests in the Greater Yellowstone Ecosystem. The FWS issued a similar opinion in the Island Park, Idaho, area of the Targhee National Forest. The agency issued a series of biological opinions that allowed a developer to build a road across U.S. Forest Service property near a productive eagle nest to build a subdivision adjacent to the eagles' primary feeding area. In both cases, the FWS issued permits allowing incidental "takings" of eagles that might occur because of human disturbances.

These kinds of problems worry many biologists when considering the effects of downlisting. In Montana, for example, half of the eagle nests are on private land. Moreover, most eagles still carry traces of pesticides in their bloodstreams, and old eagles—some live to be more than twenty-five years old—still have high levels of DDT and are not very productive. Despite the recovery successes, the eagle is still not well distributed in the West. In some areas, eagles just aren't producing young very well.

166

Successful management of fisheries in Yellowstone National Park and in blue-ribbon trout streams and lakes across the West has helped increase the eagles' prime food sources. These new sources made up for the loss of traditional food sources, such as carrion on the Great Plains and the huge natural salmon runs that probably attracted eagles from hundreds of miles to Idaho, Washington and Oregon rivers. Yet in some areas, localized declines in fisheries have led to major changes in the feeding and migration habits of entire eagle populations. The most disturbing and well-known case is the story of McDonald Creek in Glacier National Park in northwest Montana.

Every fall since 1939, bald eagles from as far north as Canada's Northwest Territories, south from Utah and Nevada and west from Oregon would converge on McDonald Creek to feast on spawning landlocked kokanee salmon planted in McDonald Lake. More than 600 eagles would return annually to McDonald Creek. The spectacle also attracted up to 30,000 tourists a year to the park to be bedazzled by the concentration of eagles. Suddenly, in 1987, the kokanee population crashed. Instead of hundreds of thousands of kokanee running up the creek, only 300 made the trip that year. Experts surmised that tiny freshwater shrimp introduced into the Flathead River system to serve as food for the salmon actually became competitors with the salmon fry for microscopic zooplankton. The change in the lake's ecosystem— the loss of kokanee—damaged the eagles' food source and the eagles left. The number of returning eagles dropped to 47 in 1987 and has not recovered since. Scientists don't know where they go. Many concentrate on the Missouri River near Helena, Montana, feeding on brown trout and other fish, but not in the numbers seen at McDonald.

Riley McClelland, a raptor biologist from the University of Montana in Missoula who has studied the McDonald Creek eagles since 1965, worried that the loss of the feeding site could spell death for many juvenile eagles. He said if salmon were unavailable at McDonald Creek, especially during autumns of severe weather and food shortages, many juveniles might never reach suitable wintering grounds and could die during the winter.

McClelland wants the FWS to slow its downlisting process for eagles. But he is only one voice, and the pressure to allow more flexibility is strong. The FWS likely will downlist the eagle in this decade.

A large contingent of federal land managers, biologists, ranchers, loggers and developers also would like to see the restrictions protecting grizzly bears lifted in the 1990s. The pressure to delist is strongest in the case of the grizzly in the Northern Continental Divide Ecosystem in northern Montana. The 5-million-acre recovery area, which encompasses Glacier National Park and the Bob Marshall Wilderness Area, has the healthiest population of grizzlies of any of the six recovery zones in the United States.

Most of the pressure to delist is coming from the state of Montana. The Montana Department of Fish, Wildlife and Parks simply believes it can do a better job of managing the state's wildlife than the federal government can. It wants to remove grizzlies from the Endangered Species List for the same reason that it supports removing endangered species protection from wolves outside the recovery area around Glacier National Park. As an agency it prefers to be in control.

On the other side of the controversy are environmentalists such as Lance Olson of the Great Bear Foundation and Keith Hammer of the Swan View Coalition, who believe the grizzly bear needs more protection, not less, to survive. In the middle is Chris Servheen, the FWS's grizzly bear recovery coordinator; it depends on when and where you catch him as to his view of delisting. In a published interview in 1989, Servheen predicted that the Yellowstone grizzly would be delisted within a decade. In 1992, however, he said there's no formal proposal to delist the Yellowstone bear, and none is likely for several years. In 1989, when Lorraine Mintzmyer, then National Park Service regional director in Denver, took over as chairwoman of the Interagency Grizzly Bear Committee, she said she wanted the grizzly in the Northern Continental Divide Ecosystem to be delisted before she left. Servheen told the *Wall Street Journal* how he felt about that prospect. "It scares me a lot," he said.

Mintzmyer was not able to reach her goal of delisting before leaving her committee chair, but environmentalists such as Hammer are convinced the drive to delist goes on, despite scientific evidence that the population is even worse off today than it was in 1975, when it was first listed as threatened. Hammer says the agencies continue to change the monitoring rules so that fewer bear sightings are mathematically manipulated to show more bears. Grizzly sightings by loggers, for example, are now considered valid and are added into the equation for meeting the original recovery goal of 440 bears. The new formula assumes that only 60 percent of female grizzlies with cubs are seen. That formula would show the bear population at the recovered level of 440 when a 100 percent sighting formula would show only 264 bears, Hammer says.

In 1989, the Interagency Grizzly Bear Committee appointed a sub-committee to write a draft conservation strategy to protect bears after delisting. Completion of such a strategy is one of the requirements for delisting any endangered species; adequate conservation measures must be in place to ensure the species does not fall back to possible extinction.

Hammer acknowledges he got a copy of the draft strategy when either he or an associate stole it from Servheen's office in 1990. But the draft, however he got it, shows the real flaw of delisting in the case of grizzly bears, whose habitat requirements are so large. The 1989 document offers a glimpse of how the delisted grizzly would be managed. The strategy is based on the idea of maintaining a "large and healthy population" of 440 grizzlies in the 5-million-acre ecosystem and outlines the protections that would be in place to preserve the bear, in hopes of keeping it from being relisted as a threatened species.

In its most revealing section, the draft strategy explains how the Endangered Species Act's all-important consultation process would be replaced. The Montana Department of Fish, Wildlife and Parks would replace the FWS in consultations, but would have limited power.

"Montana Department of Fish, Wildlife and Parks will not have the authority to prevent another agency from approving an action regardless of effects on grizzly bears or their habitat," the conservation strategy states. "The principal purpose of these consultations is to ensure that the most adequate mitigations of adverse effects have been developed and will be implemented for any proposal affecting bears."

If the grizzly is delisted in the Northern Continental Divide Ecosystem, it would be up to a hodgepodge of government agencies to preserve the great bear. The National Park Service, the Bureau of Land Management, five national forests, two Indian reservations and two Montana state agencies would have a hand in habitat management. Managers of these various agencies would meet at least semiannually to discuss grizzly management, according to the conservation strategy. Unfortunately for the grizzly bear, this committee would have no apparent legal means to stop habitat destruction.

"The committee itself does not have authority, except that members of the committee bring their authorities to the committee by their participation," the conservation strategy states. Instead, it pins its hopes on keeping the agencies working together on behalf of the grizzly: "There is a need to continue the cooperative working relationships and integrated management that developed during the time that the grizzly bear has been federally listed."

Montana's wildlife establishment has always been vulnerable to political pressure. While its staff is professional and strongly advocates protection, they admit they don't even have the funding base to protect the bear. Glenn Erickson, its representative on the Interagency Grizzly Bear Committee, says it can't afford to pay for the management alone. The state not only wants the bear delisted, it also wants the federal government to help pay for management after it gives up its authority. "We would not support delisting unless we had the funding," he says. Montana wants control but wants the federal government to foot the bill.

While dozens of endangered species don't have recovery plans at all, the FWS finished its second draft plan for grizzly bears in 1992—a revision of its first recovery plan, issued in 1982. Montana's reluctance to pick up the tab for grizzly recovery is understandable. In the new recovery plan, the FWS estimates it will cost $26 million over twenty years to recover the threatened grizzly. Because the grizzly is such a high-profile animal, there's a strong likelihood that eventually that money will be spent. Many other species even more threatened with extinction than the grizzly don't have the same prospects as far as recovery funding.

For endangered species, it pays to be popular. Twelve species, or about 5 percent of all endangered species, receive 50 percent of the federal government's funds dedicated to endangered species. Only six of those twelve are considered by the FWS to be highly threatened, and two are classified as facing only low threats to their survival. Together with another sixty species, they account for 90 percent of all endangered species funding.

In 1991, the FWS stated that the populations of species with recovery plans were more likely to grow than those without. About 60

170

percent of endangered species have no recovery plans at all, although the FWS made substantial progress in the last decade, increasing the completed recovery plans it has from only 8 percent in 1979 to 56 percent in 1987. It is the one place in endangered species management that the Reagan-Bush administrations can justifiably boast that they have improved over past administrations. There are thirteen species for which the FWS does not intend to prepare recovery plans, because they are either already extinct or recovery is progressing without them.

The cost of recovering 70 percent of all endangered species would be $770 million, said Department of the Interior inspector general auditors in a 1990 report. Additionally, using the FWS's high-range estimate of $2 million per species, recovering the 600 priority-one candidate species—those unlisted species identified by the agency as most probable to warrant listing—would cost another $1.2 billion, the auditors said. After adding another 1300 candidate species—those that biologists said eventually could be placed on the threatened or endangered species list—the potential recovery cost could rise to $4.6 billion, according to the inspector general's report.

John Turner says those numbers are deceptive because recovery costs can vary widely among species. But it is clear that at current funding levels it will be a generation before the federal government gets control of the species already facing the peril of extinction. And with global warming and other climatic trends portending even greater ecological changes, the United States, perhaps the world leader in endangered species protection, may be buried in a sea of listing petitions, recovery plans and social costs that overwhelm the institutions organized to prevent extinctions.

The budgets for the two agencies that manage endangered species have risen sharply since 1988. The U.S. Fish and Wildlife Service, which has responsibility for 638 plants and animals protected under the act, had a budget of only $18.8 million for endangered species in 1988. That rose to $42.3 million in 1992. The National Marine Fisheries Service, managing only 19 species, saw its budget go from $3.7 million in 1988 to $8.2 million in 1992. With salmon protection now on its plate, the NMFS is going to need much more. The species with the greatest funding include the bald eagle, the grizzly, the red cockaded woodpecker, the peregrine falcon, the northern spotted owl, Snake River salmon, the gray wolf and the whooping crane.

The FWS and the NMFS are not the only federal agencies that spend money on endangered species. Over the years, the Forest Service and the National Park Service have spent millions on grizzly bears. The spotted owl is costing both the Forest Service and the Bureau of Land Management millions of dollars in direct management administrative costs. Salmon restoration will be paid for mostly by the Bonneville Power Administration. In addition, states and other federal agencies all contribute administrative and capital expenditures toward endangered species protection.

171

Measuring endangered species protection simply in terms of administrative costs misses economic factors on both ends of the scale. The social costs of protection, a popular subject since the spotted owl controversy arose, have been discussed repeatedly. There also are costs to society from extinction.

The cost of extinction of the yew tree, for example, might have been the lives of thousands of women with ovarian cancer. In 1967, the National Cancer Institute sponsored a large-scale plant-screening program and found that a crude extract of the Pacific yew tree's bark killed leukemia in mice. Researchers isolated the principal active ingredient and called it taxol. Initial trials of taxol found that 30 percent of previously treated patients with ovarian cancer responded positively to taxol treatment. Cancer scientists also believe that taxol may be useful in treating breast cancer, some forms of lung cancer, head and neck carcinoma and malignant melanoma.

Scientists first depended on a natural supply of yew trees to meet the growing demand of cancer researchers for the new drug. Since it took a lot of bark to produce the drug—approximately 20,000 pounds of bark from 3000 trees to produce one kilogram—foresters searched through their forests to find healthy stands of the trees. In the Nez Perce National Forest of Idaho, seventy-five to ninety-five workers gathered bark in 1992. They were not alone. Tree poachers destroyed many trees throughout the Pacific Northwest in attempts to gather bark for sale.

Then, in January 1993, Bristol-Myers Squibb, the major supplier of taxol for cancer treatment, announced that its scientists had discovered a practical method of synthesizing taxol artificially. Zola Horvitz, Bristol-Myers Squibb vice-president, told a House subcommittee that the taxol supply problem had been solved and that the company no longer planned to harvest yew bark from federal lands. But in the short term—at least through 1997—some yew bark will continue to be harvested for taxol on private and state lands, federal officials predicted in an Associated Press article of February 4, 1993.

The Pacific yew occurs predominantly in the ancient forests of the Pacific Northwest. It grows in the understory of the old-growth forests of the Cascade Range and northern Rockies, a range that extends from southeast Alaska down to northern California. Once widely distributed, the Pacific yew is one of the victims of the liquidation of the ancient forests. Like so many imperiled species, the yew tree's decline was not a result of planned decimation but a result of neglect. As the Douglas fir overstory was harvested and run through sawmills, the yew trees were simply slashed and burned. They were viewed officially by the Forest Service as a weed species, not worth its concern.

Yew trees are gymnosperms, as are its other evergreen relatives, pines and fir. They grow twenty to forty feet high and bush out like a shrub. Their reddish-purple bark is thin and easily scales off the tree in irregular patches. The vulnerability of its bark makes the yew's survival relatively precarious by tree standards. It doesn't survive fire nor overbrowsing by animals. The common practice of clearcutting and then burning a timber stand simply destroys those yew that might have survived the removal of the overstory.

In 1990, environmentalists and the American Cancer Society had petitioned the Fish and Wildlife Service to list the yew tree as a threatened species. The petition was denied in 1991, but the message that the nation's yew resources in the Pacific Northwest must be protected hit its mark. The Forest Service immediately began an extensive yew tree inventory and started working with cancer researchers to allow reasonable harvest for research, quickly leading to the breakthrough on synthetic taxol.

The obvious lesson of the yew tree is that rare species can have values that could be lost if the species are lost. The occurrence of yew trees in generally the same habitat as northern spotted owls also helps make the case that protecting one species and its ecosystem can have benefits in the future unknown to us today. How many other wonder drugs might be growing in the wild ecosystems of the Pacific Northwest and northern Rockies?

The case of the yew tree also illustrates why endangered plants are often in a different situation than endangered animals. First and most significantly, plants, unlike wildlife, are not considered public property. Their ownership is tied to the ownership of the land. If a private property owner wants to destroy an endangered plant, he has every right to do so.

However, plants don't usually present the same land management conflicts that wildlife species do. After all, plants can't walk 125 miles in a day, like a gray wolf can. Rare plants are easier to protect because they also can be replanted and cultivated, in many cases even commercially, like the yew tree.

Yet the same considerations for protecting the genetic diversity of animals apply to plants. Simply saving individual species without preserving the genetic mix evolved over the wide area of its range could result in the loss of characteristics that make the species unique. There are seven known species of yew trees, for example, but only the Pacific yew has taxol.

Critics of the Endangered Species Act, such as Charles C. Mann and Mark Plummer, who wrote a controversial article in the January 1992 issue of the *Atlantic Monthly*, say that its major flaw is its blanket protection of all endangered species, whether or not their worth to humans or nature can be determined. Mann and Plummer stated that, whether admitting it or not, the United States government is making choices about which species to protect or not in its budget decisions and in its reactions to those asking to protect specific species. "The choice is inescapable—but the Endangered Species Act, in its insistence that we save every species, implicitly rejects this responsibility," they wrote. "As a result, the government is left with little guidance. It moves almost at random, with dismaying consequences."

Mann and Plummer don't offer an alternative to the current system, which depends on a combination of politics, priority ratings and pizzazz to decide which species get the most attention. They simply say the system should be changed because it is intellectually dishonest to say that we will protect all the species when clearly we will not.

Unfortunately, moving from the philosophy that all species have a right to exist—the so-called Noah Principle, named by biologist David Ehrenfeld—to some new, intellectually consistent position means scrapping the Endangered Species Act, the strongest protection afforded disappearing species. Starting over and devising a new Endangered Species Act as strong as the 1973 law, even if not so broad, appears to have about as much chance of surviving congressional committees as a wolf has on a Montana sheep ranch. When the U.S. Supreme Court ruled in 1978 on the Tellico Dam decision, it said the language of the act "shows clearly that Congress viewed the value of endangered species 'incalculable.' " Whether Congress meant to make such a strong stand is arguable. Yet as the case of the yew tree demonstrates, humans cannot definitively determine the value of species to future generations of humans.

174

More important, the cost to society of the current system of protecting species is not overwhelmingly burdensome, as the cases of the peregrine falcon, bald eagle, yew tree and even grizzly bear demonstrate. Only rarely does great economic disruption take place, as has happened with the northern spotted owl. Even the pain caused to loggers and rural communities in Oregon and Washington could have been avoided had the situation been addressed before it reached crisis stage. As Aldo Leopold said in 1953, "Who but a fool would discard seemingly useless parts?" And who but a nation of fools would discard the Endangered Species Act?

8. The Rocky Return of the Gray Wolf

BILL ENGET knows about the Endangered Species Act firsthand. When he says the act threatens his existence as a rancher, he speaks from experience.

For most of the year, Enget's sheep graze on public lands in Idaho just west of Yellowstone National Park. Herders push the sheep slowly on a 100-mile circuit from the lambing sheds near St. Anthony, Idaho, north to the Centennial Mountains on the Wyoming-Montana border and south in the fall to winter range on the desert near Terreton, Idaho.

Until 1984, Enget ranged his 3000 ewes and their lambs on Two Top Mountain, only a couple of miles west of the park, next to the 1600-acre ranch his family has operated since 1898. The 10,000-acre Two Top grazing allotment, with many water sources, was perfect for ranging sheep. Enget's herders never had to haul water, so they could spread the sheep widely across the mountain and meadows. It was convenient, too, since his ranch was adjacent to the allotment. If the herders were having problems with, for example, predators, Enget could be there in minutes.

Enget's ranching world changed in 1984, the year he shot a grizzly bear. In August 1983, a grizzly sow had raided Enget's sheep, killing twenty-five before moving on. Under guidelines established to protect grizzly bears, the U.S. Forest Service, which administered the Two Top allotment, should have moved Enget's sheep off the allotment after the incident. But Enget, with support from friends as high up as Senator James McClure (R-Idaho), was allowed to stay. John Burns, the Targhee National Forest supervisor, had convinced his superiors that it was better to stretch the guidelines and be flexible. He believed eventually the sheep ranchers would see for themselves that grazing sheep in grizzly habitat was not profitable. So it was the bear and her two yearlings that were moved, after biologists placed radio collars on them.

176

In August 1984, like clockwork, the grizzly sow returned within days of her arrival the year before. Again the Forest Service decided to move the bear, not the sheep. Federal bear biologists, government trappers and Idaho Fish and Game staff combed the allotment trying to capture the sow and move her away from Enget's sheep. Meanwhile, Enget and his herders worked around the clock protecting his sheep, using noisemakers to scare off the bear and moving the sheep onto his ranch.

In the early morning of August 30, Enget and his two nephews, Hal and Jeff Buster, heard a bear attacking his rams on the ranch proper. Enget, wearing only his undershorts, ran outside with his 30-40 Winchester. Hoping to frighten the animal, he fired up into the air, hitting an electrical transformer. It didn't stop the grizzly bear, which then charged the three men, according to Enget's account. Enget shot and wounded the bear, and it ran into the woods.

Few things hit as squarely at the heart of western values as the responsibility to take care of livestock. Enget says that sitting by and watching a bear eat a calf or a sheep is to a rancher what watching a child die in a fire would be to a parent. "What would my father say if he knew that I didn't take care of the sheep?" Enget asked rhetorically as he explained the incident.

Western tradition called for keeping the incident quiet. But Enget had a forest full of wildlife agents trying to trap a bear in his backyard. He also had his pride and was ready to face up to having committed the act he felt he could defend on any level. He called state authorities.

"That warden read me my rights after I told them what happened," Enget said. "I told them to go to hell."

No charges were filed, but the shooting swung the political pendulum toward environmentalists and Enget was forced to leave his cherished Two Top allotment. The Forest Service reassigned his sheep to a drier allotment in the Centennials, giving Enget only a few thousand dollars to compensate for the cost of moving them.

Federal officials finally caught the grizzly sow, which had not been the bear chewing on Enget's rams. They inadvertently killed her during the attempt to move her to Yellowstone National Park. What officials had tried to make a win-win situation had turned out lose-lose for both the bear and the rancher.

Today, Enget raises cattle on his Two Top Ranch. Grizzly bears   177 rarely attack cows, and Enget hopes that someday, when the bear is no longer listed as threatened, he can convince authorities to let him run sheep again on the Two Top allotment. Now federal officials want to reintroduce gray wolves into the Greater Yellowstone Ecosystem. For Enget that means going through the Two Top saga all over again sometime down the road. Instead of grizzles getting into his sheep, it could be wolves raiding his cattle.

"This darned wolf thing is going to be the same," Enget said. "I can deal with the wolves if they'll let me; it's the law that's the problem."

The law is the Endangered Species Act. And perhaps nothing has raised the hair on the back of more sun-tanned, leathery necks in the West than the proposal to reintroduce wolves into Yellowstone and central Idaho under provisions of the act. The descendants of the pioneers who eradicated the wolves in the first third of the century are dumbfounded and angry that anyone would want to reverse what to them was a major victory in the battle to tame the land. The Two Top saga, ending in the deaths of two bears, was repeated several times in the late 1980s. The only differences were that the predators were wolves instead of grizzly bears and the state was Montana instead of Idaho. Each time an endangered or threatened species tangled with a rancher it was the species that lost the most: its life.

Wolves, like grizzly bears, were competitors for space and control as pioneering ranch and farm families populated the West. The only difference is that grizzly bears survived, leaving them with a political toehold wolves didn't share. Grizzlies have been given their space, so to speak, even if it is too little. If wolves were going to roam the West again, a means would have to be found to shoehorn them around the region's new inhabitants.

The campaign to rid the West of wolves was carried out with ruthless efficiency. In the earliest days of cattle ranching in the West, the 1860s through the 1870s, there were few reports of cattle de-predations. Buffalo hunters were leaving thousands of carcasses across the plains, providing wolves with a bountiful food supply. Also, ranchers kept track only of calves branded in the spring, so losses may have gone unnoticed. Despite the lack of depredations, predator bounties were enacted in the late 1800s. In Montana, where the bounty was established in 1883, hunters received one dollar for every wolf pelt taken.

178   As fences began to divide up the open range, ranchers started to pay more attention to their livestock losses. At about the same time, there was a dwindling of the populations of animals wolves naturally preyed upon: buffalo had been driven nearly to extinction, and deer and elk numbers had been severely cut by market hunting. Wolves were clearly identified as the enemy by ranchers, and they, and the governments they controlled, set out to destroy them.

Those wolf warriors poisoned wolves, shot them, dragged their pups from dens and even used biological warfare against them. In 1905, state veterinarians introduced sarcoptic mange into the wolf population to weaken and kill it. Montana records show that from 1883 to 1918, some 80,730 wolves were killed for the bounties. This remarkable number may have included some coyotes, but it demonstrates just how many wolves the region supported prior to the entry of white settlers.

The drive to eliminate wolves from the northern Rockies was relentless and even included the National Park Service. Wolves had been nearly exterminated in Yellowstone by 1880, but as the wolf killing escalated in neighboring states, Yellowstone's wolf population rose by 1914. Park officials considered them "a decided menace to the herds of elk, deer, mountain sheep and antelope" in the park, and they joined the wolf-killing frenzy, even over the objections of preservationist voices both within and outside the Park Service. "It is evident that the work of controlling these animals must be vigorously prosecuted by the most effective means available whether or not this meets with the approval of certain game conservationists," a park official wrote in Yellowstone's monthly superintendent's report in May 1922.

In Yellowstone alone, from 1914 to 1926, some 136 wolves, including 80 pups, were trapped, shot or poisoned. By 1926, wolves were officially eradicated in Montana, and by the 1930s, few if any wolves roamed through Idaho and Wyoming.

Wolves continued to flourish in Canada, although ranchers and provincial governments had successfully eliminated them in the southern parts of the provinces of British Columbia and Alberta. In the late 1960s, the Canadian government decided to allow wolf populations to increase in southeast British Columbia. The government reduced hunter and trapping harvests, and the numbers rose. The wolf's tendency to disperse and its amazing adaptive characteristics took over. The wolf population began moving south.

Wolves are social species. They live and breed in packs that range in size from two to more than twenty-five, depending on the prey base and territory. The pack always consists of a breeding male and female, called the alpha pair, and their offspring. Wolves also are very territorial, establishing territories from 48 square miles to more than 900 square miles. They mark their territories by howling and by urinating on the boundaries.

179

Wolves mate in February and have their litters about sixty-three days later. Pups are born in dens in early spring and then moved to "rendezvous sites" six to ten weeks later. Those sites later become the gathering places for the pack during the summer. From there they hunt mostly deer, elk, moose or whatever hoofed mammal lives in the immediate area. They will take birds and smaller mammals, but contrary to Farley Mowat's premise in the book *Never Cry Wolf*, smaller mammals don't play a major role in their prey base.

Wolves have an incredibly high potential to increase their numbers. Wolf litter sizes range from one to eleven pups. Canadian scientists have documented summer population increases of 60 percent over the prebreeding winter population in Alberta. When a wolf pack gets larger than its range allows, young wolves often leave and look for their own mates so that they can start their own packs. Those dispersing wolves, often called lone wolves, may have been the subjects of hundreds of sightings in the northern Rockies after wolves were eradicated. Even if packs did not live in the area, those lone wolves may have visited it from Canada.

As the number of wolves in southern Canada increased, the number of dispersing lone wolves also increased. Wolves don't have rigid habitat requirements as grizzly bears do. All they need is a good prey base, suitable denning and rendezvous sites, travel corridors and for humans not to kill them. Since wolves are very prolific, the number of wolf sightings in Montana and Idaho began to rise. In 1964, 1968, 1972, 1974, 1977 and 1979, dispersing wolves were killed in Montana.

Dr. Robert Ream of the University of Montana started the Wolf Ecology Project in 1972 to investigate wolf sightings. Ream's work, based on the studies of wolves in Minnesota and Isle Royale by Dr. David Mech, laid the groundwork for future wolf studies in the West. Ream developed a reporting system for recording and analyzing wolf sightings and set the stage for ecological studies if wolves could be found. In 1979, Ream captured a lone female in the Flathead River Valley in British Columbia about five miles north of the U.S. border. The female was radio-collared and studied intensively for two years. Frustrating for Ream and his team, no other wolf was detected in the drainage during the period.

Then in February 1982, tracks of a pair of wolves were found in the Flathead River drainage in Glacier National Park in Montana and followed north into Canada. It was not known if one was the same female. Later that spring a litter of seven pups was born a short distance north of the border. The chance meeting between a lone female and male dramatically increased the chances for natural recovery of wolves in Montana. It was like magic; hence the name given the pack by Ream's team was the Magic Pack.

The Magic Pack of eight to ten wolves roamed along the border in the Flathead River drainage through 1987, with its movements watched closely by Ream and his team. In 1985, the alpha female was radio-collared four miles north of the border and seen in May with seven pups. That winter, the Magic Pack lived in Glacier National Park, and in the spring of 1986, the female denned in the park just north of Polebridge, Montana. She had five pups, the first documented evidence of gray wolves born in the western United States in fifty years.

For Ream and his team it had all been good news up to 1986. Then in March 1987, a cow was killed on the Blackfeet Indian Reservation east of Glacier National Park and the signs pointed to a wolf depredation. In May, rancher Dan Geer watched two wolves bring down and kill a steer twenty miles north of Browning, Montana. In July, two wolves were captured near Geer's 2500-acre ranch six miles from Glacier National Park. One was missing part of a leg and was sent to a wolf-holding facility in Minnesota. The other was radio-collared, released and monitored. Later that month, two wolves tried to kill a cow in the same field. Five other cattle killings took place in a ten-day period in August, sparking the traditional retribution for the wolves. The U.S. Fish and Wildlife Service (FWS) had a full-scale controversy on its hands.

All of the worst fears of livestock owners appeared to be substantiated. On August 21, the radio-collared wolf again was seen leaving a pasture where a yearling steer had been freshly killed, the third in two days. FWS staff killed the wolf, adding a new milestone to wolf recovery in Montana. It was the first government predator-control killing of a wolf in the northern Rockies in fifty-five years. Eventually, six wolves were either killed or transferred.

181

While all this was happening, the FWS was attempting to revise and update a wolf recovery plan first approved in the 1970s. In August 1987, at the time of the Montana depredations, a team of scientists from several federal and state agencies, conservation group representatives and livestock organization lobbyists released the plan they had written. It set aside three areas for recovery: northwestern Montana, central Idaho and Yellowstone. It set a goal of ten breeding pairs in each recovery zone and called for the reintroduction of wolves into Yellowstone and central Idaho.

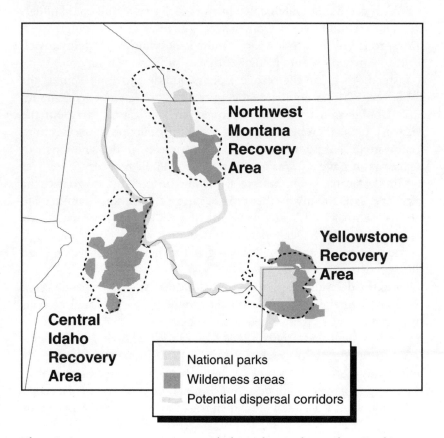

Three main recovery areas were set aside for wolves in the northern Rockies.

Frank Dunkle, former director of the Montana Department of Fish, Wildlife and Parks, mining industry lobbyist and former state senator, was then FWS director and had the power to sign or reject the plan. Although Dunkle's job was to protect the nation's wildlife resources, to be their advocate within the government, the former chairman of the Montana Republican Party was clearly worried more about ranchers than he was about restoring wolves to the West. In a speech to the Wyoming Woolgrowers Association in 1987, he said the only wolves he'd bring to the state were on his tie and that the reintroduction of wolves was an idea cooked up by the Park Service to "undo the disastrous conditions" created by the agency's mismanagement of elk and bison herds. Andrew Melnykovich, then the Casper, Wyoming, *Star-Tribune*'s Washington, D.C., correspondent, said in a November 23, 1987, article that Dunkle's behavior was such that a wolf could understand it. "If Dunkle were a wolf, he would have licked the Woolgrowers' noses, rolled over onto his back, and peed on himself," Melnykovich wrote. "Any wolf would recognize that as submissive behavior."

182

Instead of signing the recovery plan, Dunkle had John Spinks, the FWS Rocky Mountain regional deputy director, sign it. Then a month later Dunkle said he wouldn't request any budget to carry out the recovery plan, thereby preventing the reintroduction of wolves into Yellowstone. He pointed to the depredations near Browning as evidence that his agency was not prepared to handle recovery.

"Until we can get some answers, it would be foolhardy to go with the recovery plan, and my mother did not raise a foolish son," Dunkle said in a November 1987 interview with the Kalispell, Montana, *Daily Inter Lake*. "We haven't demonstrated that we are outstanding at capturing wolves, and I don't apologize for that. We just haven't had the experience."

Dunkle's decision started a political battle that has continued into the 1990s. At the center of the issue is whether wolves will be reintroduced into the Greater Yellowstone Ecosystem. The real question, however, is will westerners allow wolves to live as their neighbors?

When the debate opened in the early 1980s, ranchers were nearly universally against the return of the wolf to the West. But as wolves began repopulating northwestern Montana, their position softened, if only slightly. At least a few stockmen began to see the pawprints on the highway. Wolves were going to return whether they liked it or not, so the time to deal was now, before they arrived.

The first political leader to push this idea was Idaho's Senator Mc-
Clure, who knew the workings of the Endangered Species Act as well
as any environmentalist. It was February 1986, and Republicans con-   183
trolled the U.S. Senate. McClure was chairman of the Energy and
Natural Resources Committee, the panel with primary jurisdiction
over the Department of the Interior. He had in his hands the power to
steer the debate, and he wanted to resolve the issue before that power
slipped away.

"Wolves are coming back fast," McClure told ranchers in Idaho.
"It's inevitable. They're already in Glacier National Park."

McClure's plan was to reintroduce wolves into Yellowstone and
provide strict protection for them inside the three recovery zones. But
outside the recovery zones, the wolf would be removed from endangered status and the states would take over management.

McClure's willingness to reintroduce wolves into Yellowstone was
intoxicating to some environmentalists. Hank Fischer, the Northern
Rockies regional director of the Defenders of Wildlife in Missoula,
Montana, embraced McClure's plan as a good starting point. "Mc-
Clure clearly supports reintroduction of the wolf population in Yel-
lowstone and is out on point on this issue," Fischer told the Idaho
Conservation League in May 1986. "McClure has allowed discussions
to move forward and has encouraged talks with the livestock people.
We wouldn't even be having these discussions if he weren't leading the
way."

Fischer found himself in a lonely position. Most other western environmentalists were skeptical of McClure's plan and intentions. Historically, McClure had shown an ability to sneak language into a bill
otherwise supported by environmentalists that would aid his
commodity-using constituency. They simply didn't trust him and gave
the bill little support. But the conservation community was the least of
McClure's problems. Livestock owners' organizations, such as the
National Woolgrowers Association, were not as convinced of the inevitability of wolf recovery as McClure. They believed Dunkle and the
rest of the Reagan administration could stall off environmentalists for
years. They had the strong support of McClure's colleagues—Senator
Alan Simpson, Senator Malcolm Wallop and Representative Richard
Cheney—in the powerful Republican Wyoming congressional delega-
tion. The three opposed all efforts to place wolves in Yellowstone; even
McClure couldn't change their minds.

In November 1986, Democrats regained control of the Senate, and McClure stepped down as committee chairman. He continued to push

184 his plan, but it would be four years before he got somewhere.

Even though Dunkle and most of the rest of the Reagan administration opposed reintroduction, one voice inside Interior unabashedly pushed for it: William Penn Mott, National Park Service (NPS) director. Mott, a contemporary of President Reagan and former director of the California Parks Department, told whoever would listen that wolves belonged in Yellowstone. His argument in support of reintroduction mirrored that of conservationists: wolves are an important missing link in the Greater Yellowstone Ecosystem and should be returned to restore the balance. "Gray wolves are a part of this whole ecological system," Mott said. "I think that reintroduction is something that is desirable." He also made an appeal for the aesthetic values. "Wouldn't it be great to hear a wolf howl again in Yellowstone?" he told Yellowstone park rangers in 1987.

Yellowstone is managed under the philosophy of "natural regulation"—that is, that nature, not humans, should be the controlling power. Restoring the wolf would return to the park the one mammal that is missing from its historical mix of animals. The wolves also would help to keep down naturally the park's high numbers of elk and bison, which in the absence of natural predators such as wolves have exploded in population, causing all kinds of management problems. Renee Askins, executive director of the Wolf Fund, based in Moose, Wyoming, is perhaps the most effective voice of this view.

Askins has become the wolf's best-known defender. A thirty-three-year-old biologist with a master's degree from Yale, she is a powerful speaker whose appeal has landed her in the pages of *Time, Life, People, Parade* and on several national television programs about wolves. While she hates the *People* magazine approach that much of the national media take to stories about her and her crusade, she recognizes that it brings more attention to the issue. Askins' willingness to take her pro-wolf stand into ranch communities and other hostile environments also has won her grudging respect from wolf reintroduction opponents as well. Her presentations have moved away from strident rhetoric to a more balanced view of the wolf as an opportunistic predator that will sometimes eat livestock but that will mostly feed on Yellowstone's large wildlife populations.

Beyond the ecological arguments for restoring wolves to the Greater

Yellowstone Ecosystem, Askins says wolves are a reflection of the wildness that led to the park's preservation in 1872. "If we can't preserve wildness in Yellowstone, where can we preserve it?" she asks.

Mott and Askins found their champion in U.S. Representative Wayne Owens, a Democrat from Salt Lake City, Utah, who in 1987 introduced a bill in the House that would force the FWS to reintroduce wolves into Yellowstone. His bill didn't go anywhere, but it did provide a legislative vehicle for environmentalists to tip the debate back in their favor. Instead of talking about cattle and sheep, the subject was the ecological value of wolves in Yellowstone. Unfortunately, Owens suffered nearly universal scorn from his western congressional colleagues for pushing bills that would make wolf reintroduction a reality.

Meanwhile, Fischer and the Defenders of Wildlife were seeking to remove the strongest argument livestock owners had against wolves— the potential of livestock losses. The depredations in the Browning area had brought home the inevitability of depredations if wolves were restored to recovery population levels. Environmentalists, and even some FWS biologists, had to eat words spoken earlier that since depredation was not a problem in Minnesota, it would not be a problem in Montana, Wyoming and Idaho.

To defuse the ranchers' primary argument against reintroduction, the Defenders established a compensation program, supported by private contributions, to pay ranchers for losses attributed to wolves. The group started by paying $2239 to Dan Geer, the rancher who had suffered the brunt of the Browning depredations. "I call this supply-side environmentalism, Fischer said. "It's simply environmentalists putting their money where their mouth is."

While pleased to get the money, Geer expressed the thoughts of many ranchers in remaining skeptical about the compensation program. Verifying that wolves are responsible for a lost calf or steer is often hard with herds spread out over public land. Like most ranchers, Geer doubted that environmentalists would continue the program after wolves were recovered. He also said that compensation should be a federal responsibility, even though most ranchers doubt that the federal government would continue a compensation program if it started one.

"I don't like the law that says a wolf can come and eat my livestock and all I can do is call the government predator-control branch," Geer said in an Associated Press article published in the *Idaho Falls Post*

Register on October 25, 1987. "It seems ridiculous the way it is. All I can do is call."

186   Like Bill Enget and Dan Geer, ranchers were not going to accept, without a fight, a plan that doesn't allow them to protect their own livestock.

With all the attention on the wolves in Montana and reintroduction in Yellowstone, the third recovery area, central Idaho, was nearly forgotten. Hunters and other backcountry travelers had reported seeing lone wolves in Idaho for decades, just as in Montana. But there never was a Wolf Ecology Project in Idaho. Idaho wolves were virtually ignored by the FWS until the early 1990s. In 1978, a hunter shot a gray wolf in the Boise National Forest near Warm Lake, on the edge of the Frank Church–River of No Return Wilderness. In the early 1980s, a wolf was photographed by an Idaho Department of Fish and Game biologist conducting surveys of winter elk and deer herds while flying over the Clearwater National Forest to the north of the central Idaho wilderness. Yet no scientific analysis of wolf sightings was done in the state until 1984. Researchers Tim Kaminski and Jerome Hansen compiled 600 unconfirmed reports of wolves in Idaho between 1974 and 1983. Of these, 238 were classified as probable. From these sightings they estimated that there might be from 17 to 40 wolves in Idaho. After a two-year effort to gather physical evidence of wolves' presence in the state, they found evidence of only one to four wolves. After analyzing all the data, they concluded that no more than 15 wolves were present in central Idaho from 1974 to 1983.

Later studies showed that the number of sightings was increasing in Idaho. From 1984 to 1989, 168 wolf sightings were determined by FWS biologists as probable. Many were concentrated near Bear Valley in central Idaho, near where the wolf was shot in 1978. Other sightings clustered in the North Fork of the Clearwater River drainage near the Montana border.

Still scientists could not provide the proof. Kaminski went to work for Representative Wayne Owens lobbying for wolf legislation in Washington, D.C. Hansen, a biologist for the Idaho Department of Fish and Game, had his hands tied by the Idaho State Legislature, which passed a law in 1988 prohibiting the department from managing wolves. The law precluded Hansen from doing all but cursory research. Unlike in Montana and Yellowstone, the FWS in Idaho had little money to study wolves.

Kaminski had helped Owens place funds for Idaho wolf research in
the federal budget. However, FWS budget writers reprogrammed the money, sending it instead to northern spotted owl research. Perhaps FWS officials were worried about raising McClure's ire. McClure's wolf reintroduction legislation was based on the premise that there were no wolves breeding in central Idaho and that three breeding pairs would have to be brought in for recovery. If there were wolf packs already in Idaho, McClure had nothing to offer environmentalists in return for taking endangered species protection off wolves outside a limited recovery zone.

Despite McClure's clout, more likely it was a matter of regional bureaucratic priorities. It was just one of the many frustrations Jay Gore had before leaving his FWS post as endangered species recovery coordinator in Idaho. Gore staffed a one-man recovery bureau assigned to three endangered species: the woodland caribou, the grizzly bear and the gray wolf. The agency also had him working on snails, northern spotted owls and whatever else popped up in the region. Across the border in Montana, a recovery coordinator worked full-time on wolves and had help from three other biologists. Montana is part of the FWS's Region 6, headquartered in Denver, the heart of the Rocky Mountains, where wolves are considered a regional priority. Idaho lies in Region 1, the home of the northern spotted owl. Wolf management in Idaho was a low priority, especially since Gore didn't have evidence of breeding wolves in Idaho. It was the classic catch-22: Gore couldn't prove there were breeding wolves without research funding, and he couldn't get research funding without breeding wolves.

What made it even harder for Gore was that he was convinced there was at least one active wolf pack roaming central Idaho. He had howled for them himself in the Bear Valley area and gotten responses. "I'm convinced, and I'll always be convinced, that had I been given a crew of two to three people to look for wolves like they had in Montana, I could've trapped wolves," Gore says.

In 1991, Idaho biologists had the same kind of year Montana biologists had in 1986. Only in Idaho it wasn't active researchers finding breeding wolves, it was the public. Early in the spring, a logger discovered an elk carcass surrounded by wolf tracks near Avery in northern

Idaho. It was near the North Fork of the Clearwater area, where wolf sightings had been clustered. Then on Memorial Day weekend,
188 campers plowing their way through snow-covered roads southwest of the Frank Church Wilderness near Bear Valley claimed to see a wolf in a place called Poker Meadows, not more than twelve miles from where a hunter shot a wolf in 1978. In June, several more witnesses claimed to see wolves chasing sandhill cranes in the same meadow. Suzanne Laverty, director of the Boise-based Wolf Recovery Foundation, recorded wolf howls from what she said appeared to be three or four wolves in the same area.

The breakthrough came on July 4, when two brothers, one an Idaho Fish and Game warden, videotaped two wolves chasing cranes in the same meadow. Wolves were traveling in a pack in Idaho, and now there was evidence.

On August 1, Jay Gore got a call from Chuck Williams, a Lowman, Idaho, cowboy who was herding cattle in Bear Valley. Williams had tried unsuccessfully to help an injured wolf, he told Forest Service staff, and wanted help from authorities. Gore and Idaho Fish and Game veterinarian David Hunter flew to the scene in a helicopter and found a two-year-old female wolf seriously injured in the head and paralyzed. The wolf was airlifted to Boise and died the next day. A necropsy later showed that the wolf had been poisoned, killed by an agricultural chemical called Ferdan. The injury, Gore surmised, probably had resulted from an encounter with an elk or a moose. The wolf's death was disappointing to Gore, but its discovery proved what he had been saying all along.

In October, Ray Lyons, the Idaho Fish and Game warden who had videotaped the wolf in July, was camping once again in the Poker Meadows area. Soon after he let his dogs out the morning of October 8, they came running back in a great commotion. Lyons looked out from the camper and saw a big wolf chasing the dogs. Alongside the timber at the edge of the meadow, three other wolves stood watching.

"I'd say there many have been a pack in the area for quite some time, even as far back as 1978," Gore says. "Ranchers in the areas have been telling me for years they saw wolves, but we couldn't confirm it until 1991."

Montana continued to be the center of wolf activity in the northern Rockies in the mid-1980s. The Magic Pack's offspring were dispersing throughout the region. However, this prolific pack faced new threats to its existence from north of the border. In September 1987, the pack had split into two separate packs: the Sage Creek Pack, made up of four 189 adults and five young, and the Camas Creek Pack, made up of four adults and six young. The Sage Creek Pack had its rendezvous area just north of the border in British Columbia; the Camas Creek pack was centered on the western side of Glacier National Park. Both packs followed the game herds north in the summer and south in the winter.

On September 10, the British Columbian provincial government opened a season on wolf hunting for the first time since the 1950s in the area just north of the border. On September 17 and 18, an adult male and female were killed legally by hunters. The female, a known member of the Sage Creek Pack, was shot in the Flathead River drainage. The male was shot in the nearby drainage of the Wigwam River. Two more wolves were wounded in the Wigwam area on September 20. One was a radio-collared male.

Then two more wolves were killed by hunters in the area before Canadian and U.S. officials met to discuss the effects of wolf hunting in Canada on wolf recovery in the United States. They signed a pact on October 23 to close the British Columbia hunting season in the southeastern corner of the province. It was not to take effect until lower-level officials signed off. Two days later, hunters shot three known members of the Sage Creek Pack in the Flathead River drainage. On October 26, the appropriate papers were signed and the hunting season was closed twenty-four hours later.

By 1989, the FWS had positively identified three packs along the border: the Camas and Wigwam packs and a third pack that had moved in, called the Headquarters Pack. Moreover, a fourth pack had established itself in the Ninemile Valley drainage, twenty-five miles northwest of the highly urbanized Missoula, Montana, area. The Ninemile wolves were the southernmost wolf pack in Montana and showed just how quickly the wolf population was moving south. The pack gave hope to those who believe it's better for wolves to recolonize the northern Rockies themselves, instead of being reintroduced by humans. The Ninemile wolf story also provides a vivid picture of the challenges faced by wolves and humans when they live side by side in the modern world.

The saga began in April 1989, more than seventy miles north of Ninemile Valley, in Pleasant Valley, west of Kalispell, Montana. A sheep rancher saw what he thought was a dog or a coyote entering his sheep pen and he shot it. The animal turned out to be a two-year-old male wolf. The FWS decided against charging the man despite the requirements of the Endangered Species Act. It wanted to be responsive to the public and held a public meeting in the area. Local ranchers told FWS officials that a pack of wolves had formed in the area and, much to their chagrin, was producing pups. This was livestock country, and the ranchers were not about to give it up. The wolves had chosen private land for their denning and rendezvous sites, and it was outside the recovery area designated for wolves. There also was evidence that the wolves had developed a taste for livestock. FWS officials decided the wolves couldn't stay.

Government trappers captured two pups, each weighing 40 pounds, an adult male weighing 100 pounds and a female weighing 78 pounds. One pup couldn't be caught. It was later suspected of killing cattle and was shot and killed by federal agents from a helicopter. The rest of the pack was radio-collared and released in Glacier National Park.

Unfortunately, the pack didn't live happily ever after. The trauma of relocation apparently is as hard or harder on wolf families than it is on human families. The adults immediately abandoned their pups, which died, and the male later died of complications from a trapping injury. The female was on her own. Ignoring the artificial boundaries that delineated the recovery area and safety, she trotted southwest out of Glacier into Swan Valley and nearly 200 miles south to Ninemile Valley. There she found a mate. Six pups were born in April 1990, establishing the Ninemile Pack.

The Ninemile Valley is a long, relatively narrow valley filled with livestock ranches. Several are large, traditional ranches where the owners' livelihood is dependent on the ranching operations. Most, though, are "blue collar" ranches, run by families whose major source of income comes from working in the paper mill in nearby Frenchtown. At first glance, the valley would appear to be a lousy place for wolves to live. Human population is relatively dense, especially on the southern end, and cattle, sheep and horses are spread throughout. The Ninemile area has been extensively clearcut during the last two decades. The surrounding mountains are a patchwork of old tamarack trees, young aspen and openings that is ideal for whitetail deer. And deer are everywhere as you drive through the valley, making nighttime driving hazardous on its dozens of miles of roads.

The drainage immediately north of Ninemile Valley is relatively wild, quiet and undisturbed, seemingly the perfect corridor for wolves moving between Montana's northwestern recovery zone and Idaho's central recovery zone only a few miles west of both valleys. So far it is missing only one thing—wolves.

Apparently, because of the presence of deer, wolves seem to prefer Ninemile Valley. The presence of wolves in the valley was unsettling to local residents and those in the surrounding countryside mainly because it showed that wolves could turn up anywhere. A month after the female whelped her six pups, she was found dead, her radio collar cut off and damaged, even though not a single incident of livestock depredation had taken place. There was no welcome mat for wolves in the Ninemile Valley. The male stayed with the pups and raised them, until he, too, succumbed to the perils of civilization. In September 1990, he was struck by a vehicle and killed on Interstate 90 just south of the valley, leaving the pups orphans—and unwelcome orphans at that. That's when Mike Jimenez entered the picture.

Jimenez, an FWS researcher and a veteran of Robert Ream's Wolf Ecology Project, found five of the pups and radio-collared two of them. It was obvious that without parents the young wolves would die. So Jimenez, the father of a teenaged son himself, became foster parent to the wolf pups.

He roamed the backroads of Montana between his home in Missoula and the Ninemile Valley gathering roadkill to feed the growing wolves. He followed them with a radio transceiver as they got to know the valley that was their home. Soon, even some of the ranchers took a paternal interest in the young wolves, which had stayed away from their livestock. Like a father, Jimenez took pride in his adolescent wolves as they learned to hunt whitetail deer on their own. It looked like Montana was going to have a wolf success story, perhaps a model to show that wolves and humans could coexist.

In December, four of the six wolves, perhaps feeling their oats, left the friendly confines of Ninemile Valley and headed northeast over the mountains to new country near the towns of Dixon and Ravalli. Jimenez got a call late one night in April 1991. A dead steer had been found on a ranch near Dixon, and the evidence pointed to wolves.

"It's like your kids being picked up for shoplifting," Jimenez says. "You say, 'It can't be my kids; they're upstairs in bed.' I told them it couldn't be my wolves."

192

Then another steer was found dead, and Jimenez had to accept the obvious: his wolves had left the valley and might well be acting like, well, wolves. He went out to check on the kill and found two of his adolescent wolves chasing their tails, playing like rowdy, pubescent youths on the loose. "Prebreeding subadult males," Jimenez quips. "It cuts across species."

Federal trappers captured three of the wolves, including the female, which was pregnant, and moved them to Glacier. Now there were two wolves left in the valley, and wolf recovery in the Ninemile was back where it had begun. Those that were moved suffered the typical fate of Montana wolves. One was found dead in Mud Lake near Bigfork, Montana, northwest of Ninemile Valley. A second was shot and killed by a Condon rancher for attacking cattle. The rancher, by the way, was not prosecuted. The third was trapped and removed permanently to a research facility when it was suspected of killing sheep near Dupuyer, Montana.

The vulnerability of wolves to human-caused mortality—either as roadkill or rifle fodder—is the chief threat to the recovery of wolves in Montana and, in fact, in the entire northern Rockies. The sad story of wolf recovery in Montana, starting with the Magic Pack and continuing down to the Ninemile Pack, is that nearly all suffer when they come into contact with humans. "They don't die of old age," says Jimenez.

Yet in 1991 the question remained: why had the wolves of Ninemile stayed away from the livestock there but attacked steers elsewhere? Jimenez remains baffled. Rancher Dick Ramberg thinks he knows.

Ramberg raises sixty cows on his 600-acre ranch about halfway up the Ninemile Valley. The older wolves came right across his ranch, he says, and didn't cause problems. It was winter then, and he had his cattle close to the house; so it was harder for the wolves to cause problems, since he was nearby to protect his animals. The young wolves had more opportunity, he says, but they simply weren't big enough.

"Most of us have cows and calves, and the old mother cows are pretty good at protecting their calves," says Ramberg. "I think that's why they didn't bother them."

The end of the Ninemile Pack was not the end of the Ninemile story. Soon after the young wolves left, a female with a radio collar moved in from Glacier, following the same route as the mother in the Ninemile Pack had. The new female and two other wolves, perhaps the remaining two of Jimenez's pack, roamed the Ninemile together and for sixteen months stayed away from the livestock. Ranchers once again became less hostile toward their wolf neighbors, accepting and not disturbing the wolves as long as they didn't bother their livestock.

Then, in April 1992, a 500-pound steer was killed on the small ranch of David Fish in the Ninemile Valley. The wolves immediately became unwelcome. Fish was compensated by the Defenders of Wildlife, who tried to calm ranchers' fears. And the cycle continues.

The conservation community had based its case for wolf recovery on two premises: that it would be necessary to reintroduce wolves into Yellowstone and central Idaho and that the economic effects of wolf reintroduction, mostly livestock depredations, would have to be controlled. Reintroduction into Yellowstone in particular had become a national cause célèbre, a Green measuring point on which politicians and bureaucrats alike were rated. In 1990, Senator McClure finally got hearings on his wolf reintroduction bill, and despite opposition from both environmentalists and livestock owners, he was close to reaching the consensus needed to move forward. But McClure had run out of time. He chose not to run for reelection, and no one carried on his crusade to trade reintroduction of wolves into Yellowstone and central Idaho for delisting them everywhere else.

Enter John Turner, FWS director beginning in 1989, Wyoming outfitter and livestock owner. Turner proposed designating reintroduced pairs of wolves in Yellowstone as an experimental population, giving managers more flexibility to kill them when they leave the national park. His plan, approved by Congress in 1990, established a Wolf Management Committee to write a compromise management plan, with representatives from federal agencies, the states of Montana, Idaho and Wyoming, environmentalists, livestock owners and outfitters. His idea sparked one representative of a livestock group to call for Turner's resignation, and he was widely criticized in Wyoming. His political mentors, Wyoming senators Malcolm Wallop and Alan Simpson, did not publicly back the plan. Turner says he could have avoided the issue, which is what his predecessor, Frank Dunkle, had done, but he wanted to resolve the issue once and for all.

"As an outfitter I understand the legitimate concerns about the game herds. Being in the cattle business I understand the legitimate concerns of stock growers who use public lands," he says. "So, politically it's probably not very smart. But Wyoming people and people in Idaho and Montana are not going to wish this problem away."

Turner believed it was inevitable that wolves would eventually return to Yellowstone on their own. Like McClure, he wanted to make the best deal possible for his region while he still had cards to play. The Wolf Management Committee, which because of the political maneuvering of western congressional delegations was stacked in favor of livestock owners, held hearings in all three states and attracted hundreds of wolf opponents and supporters in what became good theater but less effective policy-making procedure. The final plan would have permitted reintroduction into Yellowstone and central Idaho and protected the wolves in the recovery zones. It would place management outside the recovery zones in the hands of the states, which had indicated in hearings that they would allow ranchers to control the wolves themselves if the wolves attacked their livestock. Environmentalists on the panel balked when it became obvious that many of the wolves already living in the northern Rockies would become fair game. Moreover, the language of the plan required further congressional action, which sparked widespread opposition from environmental groups, which feared it as a threat to the Endangered Species Act because it removed protection for an endangered species. Turner's plan was dead on arrival in Washington; it was time to go back to the drawing board.

In 1987, the Wyoming congressional delegation had nixed any attempt by the National Park Service to write an environmental impact statement on reintroducing wolves, hoping to delay consideration of the proposal. Finally, in 1991, the FWS and the Park Service were directed by Congress to study wolf recovery and propose how to proceed.

Meanwhile, many environmentalists were becoming increasingly frustrated with the process and even with their own major advocates, such as the Defenders of Wildlife. Keith Hammer, executive director of the Swan View Coalition in Bigfork, Montana, vented his feelings in an article in the fall 1991 *Predator Project Newsletter*.

"By placing the concerns of the livestock industry above the needs of the wolf, wolf control was seen as inevitable and a compensation fund was established to assuage fears of economic losses," Hammer wrote. "These concessions became the terms of discussions behind which the vast majority of the conservation community rallied. . . . Years later we don't know of any established resident wolves in Yellowstone or

Central Idaho! Our effort to initiate a reintroduction project has thus far failed. And in the Northern Continental Divide Ecosystem, where wolves have returned on their own, our bargaining chips have gotten us at least nine dead wolves in the past 24 months, a population of no more than 50 wolves . . . and $100,000 of wolf advocates' support sitting in a bank account, ready to pay off ranchers who have lost some of their non-native commodities to natural processes. This cannot be construed as even a partial victory."

Hammer argued for scrapping the reintroduction strategy— restoring wolves more quickly by reintroducing them into Yellowstone and Idaho—and instead working to force the FWS to begin protecting the wolves they already have. His proposal was based on the idea that wolves already were recolonizing Montana and central Idaho and that it was just a matter of time before they reached Greater Yellowstone as well. "As wolves naturally recolonize their old haunts we will increasingly be given the opportunity to take the real moral or ethical high ground without any threat of losing political or geographic ground," Hammer wrote.

Conservationists have had public opinion on their side all along, even in the three northern Rockies states. A nationwide poll, including a random sample of members of the National Cattle Association, American Sheep Producers, National Trappers Association and an oversampling of Rocky Mountain state and Alaska residents, found 42 percent supporting the wolf and 30 percent hostile toward the wolf. In the Rocky Mountain region, the poll, conducted by Stephen Kellert, showed 50 percent liking the wolf and 30 percent disliking the wolf. Poll after poll in Idaho, Montana and Wyoming showed positive feelings toward wolves and support for wolf reintroduction. In 1990, Alistair Bath and Colette Phillips of the University of Calgary found that 53.3 percent of Idahoans surveyed had a positive attitude toward wolves, as did 44.7 percent of Montanans. Only 11.9 percent of Idahoans and 22.3 percent of Montanans had negative feelings toward wolves. A study conducted by Bath in Wyoming in 1987 showed that 47.2 percent of respondents had positive feelings toward wolves. A poll conducted by Boise State University political science professor John Freemuth in 1992 showed an astounding 72 percent of Idahoans favor having wolves in wilderness and roadless areas of central Idaho.

Despite the frustration of some environmentalists, reintroduction continues to be the choice of the major environmental groups. Advocates such as Hank Fischer still support a less confrontational approach, working through the conflicts rather than pushing the wolf down the throats of rural western residents. After all, Fischer says, "shoot, shovel and shut up" is still alive in the West, as several of the Montana wolf deaths show. How many more wolves have died with no trace?

"Wolf recovery will be much smoother if we can reintroduce wolves into the areas where there won't be conflicts, like Yellowstone," Fischer said.

John Turner's prediction that wolves would eventually return to Yellowstone on their own turned out to be true. With little help from environmentalists, and despite the secret killings by hostile livestock owners, wolves have moved into Greater Yellowstone on their own.

In October 1991, Dr. William Fogarty, an orthopedic surgeon from Jackson, Wyoming, and his guide, Rod Doty, were three days' ride north of Buffalo Valley on the northeast side of Jackson Hole, southeast of Yellowstone National Park. The pair were hunting bighorn sheep in the heart of the Teton Wilderness with Fogarty's son Bill, a lawyer from Cody, Wyoming. Perched on the side of a mountain above 10,000 feet, they were scanning the surrounding peaks and valleys for sheep.

"All of a sudden I saw something moving," Fogarty says. "I put the spotting scope up and saw a big gray wolf." The animal was walking above tree line, giving Fogarty and Doty a good long look. "It had the typical facial features of a wolf," he says. "Both of us agreed it was bigger than a coyote."

The next day Bill Fogarty found tracks in the same area measuring three inches in diameter. "That's one big dog; that's a wolf," Fogarty said.

Fogarty's report, first published in the *Jackson Hole Guide*, a weekly newspaper in Jackson, Wyoming, came at a time when reports of wolves in the Greater Yellowstone Ecosystem were increasing dramatically. Especially in the DuNoir Creek drainage, just east of Togwatee Pass near Dubois, Wyoming, only a few miles from where Fogarty saw his wolf, the reports of wolf sightings were becoming too numerous to ignore. A rancher told FWS biologists that he had seen a pack of wolves in the area the year before. The same month Fogarty saw the wolf, state and federal biologists investigated another report of wolves howling in the area. Several biologists reported hearing what they thought were wolves while checking out the reports. A Wyoming resident had also reported seeing a radio-collared wolf at Togwatee Pass in December 1990, though officials said they knew of no radio-collared wolves in the area. In December 1991, a couple videotaped what they said was a wolf on the east side of Yellowstone National Park. In February 1992, two Bureau of Land Management biologists said they saw a wolf near Cody.

The pattern was familiar. Like the early reports in northwest Montana and central Idaho, the signs were pointing to there being perhaps a pack or more of wolves already setting up housekeeping in Greater Yellowstone. As always, the FWS was cautious in its assessment. "We're treating each sighting the same," said Jane Roybal, of Cheyenne, the FWS biologist who is coordinating the search effort in the Yellowstone area. "Once we get a cluster of reports we go in and do field checking."

The breakthrough came on August 7, 1992. In Yellowstone National Park, Ray Paunovich, a cinematographer from Bozeman, Montana, saw what appeared to be a wolf feeding on the carcass of a bison along with two grizzly bears and a coyote. He filmed the scene, and Kevin Sanders, who was working with Paunovich, shot still photographs. Yellowstone brought in a team of top wolf researchers to examine the films, and they were in agreement that the animal looked like a wolf. But without genetic tests, possible only if the animal were caught or killed, a confirmation was impossible.

Despite this strong evidence, there remained doubters. The *Jackson Hole Guide* chided Yellowstone officials for not chasing down the animal so that indisputable proof could be found. Other critics of reintroduction suggested that the wolf was a wolf-dog hybrid, perhaps planted by overeager environmental extremists. Jerry Kysar, a hunter from Worland, Wyoming, inadvertently provided the evidence that even the *Guide* couldn't ignore. He killed a wolf he thought was a coyote in the Teton Wilderness just south of Yellowstone on September 30, 1992.

"It all took place in less than thirty seconds," Kysar told the *Billings Gazette*. "I jumped off the saddle, grabbed the gun out of the scabbard, chambered a round and shot."

The wolf was running ahead of a pack of three or four other animals. Kysar opted to turn in the wolf to authorities, facing the possible

$10,000 fine stipulated in the Endangered Species Act that had yet to be imposed on modern wolf killers. FWS researchers also saw the pack, and Ed Bangs, who heads the team conducting the environmental impact statement on wolf reintroduction, acknowledged that the 1992 wolf sightings will force the agency to consider new management alternatives. If wolves are already in Yellowstone, then it would be foolish to remove protection for existing wolves in exchange for protection of a reintroduced experimental population.

If wolves reestablish themselves in Montana, Idaho and Wyoming, only the bullets of hunters and ranchers will stop them from repopulating much of the rural West. Already, environmentalists in Colorado are calling for studies of wolf reintroduction there. With limited protection, wolves have proven that they can find a home in the small spaces left for them in between humans and their estates. They have established the beachhead from which they can grow and prosper. Ranchers like Bill Enget will still be in business, loggers will still log and miners will still mine. When wolves repopulate the northern Rockies, the Endangered Species Act will have yet another success story, and the haunting sound of howling wolves will once again echo across the landscape.

However, if the Wolf Fund's Renee Askins is right, it could be years before wolves gain a solid foothold in places such as Yellowstone without an ambitious and sustained augmentation program. The political leaders of the ranching community will continue to fight such a program with religious fervor. Only now they will no longer have allies in the Department of the Interior. Bill Clinton's election gave the gray wolf powerful friends in Washington whose presence should speed the spread of wolves southward.

9. Protecting the Wild Rockies

THE wild Rockies endure where the mountain caribou once roamed. Mountain caribou ranged up and down the northern Rockies south to Wyoming as recently as the 1800s. Today, only a remnant herd, fewer than 100 animals, lives in the Selkirk Mountains straddling the U.S.-Canadian border. The caribou, the rarest mammal in the United States, retains only a precarious hold in the Lower 48 states.

The caribou's rapid retreat north follows the pattern of large endangered species throughout the West. The grizzly bear and gray wolf once lived as far south as Mexico. Nowadays, both species inhabit an area only as far south as the caribou lived in the 1800s. They have been pushed north by civilization as if driven by native beaters on a hunt.

Hunting, poaching and habitat destruction have forced the species' retrenchment. While hunting and poaching can be controlled relatively easily, habitat destruction is hard to reverse. The continued fragmentation of the northern Rockies reached a feverish pace in the 1980s, and today the future of the mountain caribou, grizzly bear, gray wolf and hundreds of other species depends on the preservation of the remaining large tracts of undeveloped land. In the continental United States, only in the northern Rockies does such an opportunity still exist to preserve wild lands and large wild animals.

The northern Rockies bioregion includes about 50 million acres of landscape from Willmore Wilderness Park north of Jasper National Park on the British Columbia–Alberta border south nearly 1000 miles to the Greater Yellowstone Ecosystem of Wyoming, Montana and Idaho. It stretches from nearly arctic territory on the north, dominated

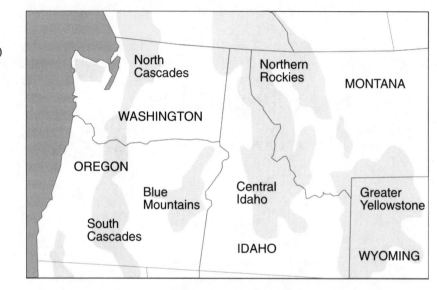

Bioregions of the Pacific Northwest. (Source: Defenders of Wildlife magazine.)

by glaciers and icefields, to the deserts of eastern Idaho and western Wyoming on the south. Despite the road building, timber cutting and residential development of the 1980s, nearly half of the region remains in a virtually wild state. In Canada, three national parks and several wilderness areas protect a core of the Rockies. In the United States, two national parks and about a dozen wilderness areas protect portions of the southern end. Moreover, many large, roadless areas remain undeveloped. Most of these de facto wilderness areas remain unprotected on both sides of the border. Even more areas only slightly developed or partially logged remain usable habitat for the large endangered mammals. These fringe areas are the most vulnerable habitat and face ecological destruction by saw and bulldozer. Only careful development, logging or land-use planning in the fringe areas can keep them available to the grizzly bear, gray wolf and caribou and permit safe travel north to Canada and south to Yellowstone. The areas that are protected, such as Glacier National Park and the Bob Marshall Wilderness Area, appear large by human geographical terms. But for large mammals that can travel up to 100 miles a day and need to be mostly isolated from human activity, even a million acres is small. The largest wilderness area and national park complex is only about 4 million acres.

Preserving the ecological integrity of the remaining land is the great challenge of this decade and the coming century. The current threat comes from traditional industries and attitudes developed for a time when the population was sparse and spread out. Today, thousands of new residents are flocking to the northern Rockies, building homes in once-isolated canyons, filling up backcountry trails on the weekends and adding new stresses to a system already on the verge of breakdown.

The experience of the mountain caribou offers a preview of what we may see for an entire community of species that now make the northern Rockies one of the last wild places left in the United States. The southern end of the northern Rockies always was only marginal habitat for mountain caribou, a subspecies of the more common woodland caribou, which lives primarily in Canada. Even when these reindeerlike ungulates with large, distinctive antlers ranged into southern Idaho and Wyoming, they occurred in small, isolated bands. The first press of white settlers into the region in the 1800s quickly drove the caribou range north, to where the population was contiguous with the larger population in Canada. This included the northern edge of Idaho, Washington and Montana but centered mostly on the Selkirk Mountains, which thrust north from Washington and Idaho into British Columbia. Caribou in the Selkirks were so common that Kootenai Indians, early miners and trappers hunted them. Market hunters and outfitters also briefly exploited the beasts.

From 1900 until 1950, caribou were seen regularly in the Selkirks in groups of one to twenty animals, but the population may never have exceeded fifty. Still, these animals lived in near isolation, since few people ventured into the high mountain forests where they lived until the 1950s. Then humans and nature combined to dramatically alter the world of the Selkirk caribou.

In 1949, severe winds blew down large blocks of spruce, which were soon infested with spruce bark beetles. The beetles spread to many of the spruce that remained standing, and much of the mature forest on which the caribou depended was destroyed. Then roads were punched into the high-elevation habitat to allow loggers to salvage the remaining spruce. In British Columbia, a major highway was built through the center of the area, and eventually power lines and pipelines followed. In the late 1960s, clearcuts and fires destroyed even more precious habitat. Civilization had caught up with the caribou once again, sending them to their deaths on the highways and at the hands of careless hunters or poachers. By 1970 the population throughout the Selkirks dropped to as low as twenty animals.

202

The demise of the mountain caribou is a typical story of the road to extinction. But thanks to new management initiatives and the federal Endangered Species Act, the final chapter on the caribou in the United States has yet to be written. In 1982, the mountain caribou was listed as an endangered species. Following the listing, an intensive management plan was written to preserve the old-growth spruce and subalpine fir the caribou prefers. Several roads were closed and a massive public education program started. Sixty caribou were transplanted into the Selkirks from other herds in British Columbia.

The augmentation and management program so far appears to be a success. Yet the current program is designed to recover caribou only in the Selkirks. Although there is evidence that caribou have moved into the Yaak Mountains in Montana, the U.S. Fish and Wildlife Service (FWS) has steadfastly refused to consider restocking that area and other former caribou range in the region. For now, if caribou are to survive in the United States, they will do so by maintaining a tenuous foothold in the Selkirks.

The caribou program covers only a relatively small area of land, yet its management scheme represents how the entire northern Rockies can be preserved without halting all development. Some of the land in the Selkirks is preserved as wilderness. Other areas are open to development, with the welfare of the caribou taken into account in any project.

Locking up the remaining threatened lands of the northern Rockies in wilderness and national parks is probably not only impossible but unnecessary for the survival of the region's large endangered species. Preserving the integrity of the natural ecosystems—whole, natural life-support systems for entire communities of plants and animals—is the key to the long-term survival of these species. It also may be critical to the long-term ecological health of the entire Pacific Northwest. Contrary to the prevailing wisdom in the West, protecting the land needed to preserve caribou, grizzlies and wolves will benefit the region's economy as well.

Preserving large endangered species also protects many plants and other animals that depend on the same ecosystem. And smaller ecosys-

tems depend on the larger regional whole. Preservation of the caribou in the Selkirks depends on preservation of the caribou throughout British Columbia and Alberta. Likewise, the fate of the grizzly bear in   203
Yellowstone is inextricably tied to the fate of the bear in Glacier and beyond.

Species diversity and habitat size are the features that contribute most to protecting the integrity of ecosystems. If an ecosystem is large enough and has all of the species and habitat it has carried through thousands of years, it can adjust to all but the most dramatic climate changes and huge catastrophic events. As the larger areas are fragmented, however, into smaller, isolated habitats, species are more likely to become extinct.

Ultimately, fragmentation of wildlife habitat threatens the genetic diversity of the species. While some argue that genetic diversity can be preserved in gene banks or zoos, these measures cannot protect the interchange of various genes that takes place in nature. It is this interchange that is at the heart of the evolutionary process.

Fragmentation of habitat leads to inbreeding, since the choice of sexual partners in an isolated population is limited. Inbreeding permits a species' weakest traits to thrive in a population. Other effects of inbreeding include lower fertility, smaller offspring and higher infant mortality.

Preservation of the large mammals of the northern Rockies has been based partly on reaching and ensuring minimum viable population sizes that contain enough animals to protect the genetic base and prevent severe environmental fluctuations from wiping out the species. In estimating a minimum viable population, the wild-card factor is the amount of risk of extinction figured into the equation. It is reflected in the percentage of risk of extinction over a given period of time. A minimum viable population that is based on a 90 percent likelihood that the species will survive for 100 years will be smaller than a minimum viable population with a 90 percent chance of survival for 300 years.

The FWS uses 100 years for the grizzly bear, for example, on the recommendation of the Interagency Grizzly Bear Committee. Based on this figure, it estimates a minimum viable population of as few as 70 to 90 bears in the Selkirk and Cabinet-Yaak ecosystems. Independent researchers estimate that a population of from 1600 to 2000 bears is necessary for bears to survive 300 years or more. Mike Bader, executive director of the Alliance for the Wild Rockies, a group based in Missoula, Montana, that seeks to protect the remaining wild lands in the region, argues that, using the higher minimum viable population estimates and the average area requirement reported by biologist Mark Shaffer of thirty-three square miles for each bear, a relatively undisturbed land mass of 35 million to 42 million acres may be needed to preserve a viable grizzly bear population. He has powerful allies in that opinion, including noted grizzly researcher John Craighead.

The presence of humans does not in itself destroy an ecosystem. Humans are a part of all the ecosystems of the northern Rockies. What sets these natural ecosystems in the northern Rockies apart from the human-dominated ecosystems of southern Idaho's agricultural region, Washington's Puget Sound or Manhattan Island, for that matter, is that these ecosystems are sustained almost wholly through natural recycling processes. Starting with the sun, energy, nutrients and water naturally work through the system to produce a sustainable environment for the life forms of the ecosystem. A human-manipulated system, however, sustains itself less through natural recycling and more through the addition of nutrients and energy from outside the system. The human-manipulated system has less diversity and therefore is less stable.

Scientists find that large, uninterrupted geographic areas not only hold more animals than smaller ones hold, but they also hold more kinds of animals. That is the heart of an ecological concept known as island biogeography. The longer an island has been isolated, the less its plants and animals have in common with their mainland counterparts, scientists have discovered. Eventually, the overall diversity of the island population drops sharply. Large animals, unable to maintain large populations in relatively small areas, begin to inbreed, thus inhibiting their abilities to adapt to changes in the environment. Scientists say the same thing can happen with isolated land ecosystems. Instead of being surrounded by water, for example, the Greater Yellowstone Ecosystem is surrounded by a sea of development, separated from other ecosystems by the roads, clearcuts and subdivisions of human society.

Ecosystems don't follow political boundaries; they set their own boundaries. Human activity can limit those boundaries, but only the movements of wild creatures can delineate them. Because animals don't follow simple patterns, ecosystem boundaries are fluid, ever-changing. An ecosystem can be as small as an elk carcass or as large as the world itself. "Everyone wants to take their magic marker and draw a line, but from a scientific standpoint you have to draw it vaguely," says John Varley, Yellowstone National Park's chief researcher. Deciding what constitutes an ecosystem for scientific purposes is only slightly more sophisticated than drawing the borders of counties, states and nations. Even though an ecosystem cannot be defined precisely, scientists can identify a geographical area that includes the vast majority of species and geophysical characteristics that set it apart from other ecosystems. Varley says he can identify the Greater Yellowstone Ecosystem, roughly 18 million acres, or 28,000 square miles, in this context. "But if you want to put up a fence that includes 100 percent of living organisms and the inorganic material, then it can't be done other than in a global context," he says.

There is some scientific agreement concerning what areas constitute relatively intact ecosystems in the northern Rockies. The ecosystems include the Selkirk Ecosystem in Idaho, Washington and British Columbia; the Cabinet-Yaak Ecosystem in Idaho, Montana, British Columbia and Alberta; the Northern Continental Divide Ecosystem in Montana and Alberta; the Central Idaho–Bitterroot (or Greater Salmon) Ecosystem in Idaho; and the Greater Yellowstone Ecosystem in Montana, Idaho and Wyoming. The major activities responsible for the fragmentation of these northern Rockies ecosystems have been timber cutting and road building. Thousands of miles of roads were built in the region, beginning in the 1940s, to open up the forests to the timber industry. The road-building and timber-harvesting spree quickened in the 1970s and peaked in the 1980s, when owners of large, private timber lands began clearcutting their holdings and the U.S. Forest Service increased its allowable harvest. In no place were the effects more evident than in Montana.

A traveler driving through Montana on the extensive interstate highway system might not notice the effects of logging, since timber cuts are rarely visible from highways. If you drive off on side roads through places such as Swan Valley or the Cabinet Mountains behind Thompson Falls, however, you will see giant holes in the forest— sometimes in perfect squares—where loggers have clearcut to the boundaries of private land. From atop the mountain peaks of the Beaverheads or the Mission Mountains the extent of the harvest is obvious, as clearcuts lie like patchwork across the mountainsides.

Montana, like the entire Pacific Northwest, has been controlled throughout its history by both economic and governmental powers outside its borders. Copper companies such as Anaconda, railroads such as Northern Pacific and later coal and oil companies invested the capital and sent the profits to Minneapolis and New York. The federal government, most notably the Forest Service, controls millions of acres of land, especially in western Montana, and its regulations and policies have reached into the daily lives of most rural residents. In the past, Montana and Idaho were virtual colonies economically to the eastern and midwestern United States. Since World War II, reliance on the traditional extractive industries of mining and wood products has declined. But the attitude of the states' political and business leaders remains rooted in the belief that the economy rises and falls with these two very cyclical industries. Consequently, the timber industry, still very much controlled by eastern business interests, has been allowed to manage Montana and Idaho lands with little interference from the states, despite the detrimental effects to the land and subsequently to the future of the communities.

In Montana, the two largest landowners and timber cutters, Champion International, which owns 1.7 million acres, and Plum Creek Timber Company, which has 825,000 acres, went on a clearcutting spree unprecedented since the turn of the century. Both companies, for entirely different reasons, made corporate decisions to liquidate their Montana trees. That meant departing from the traditional, scientific limits on cutting. Both companies began cutting trees faster than they grew, harvesting above sustained yield.

Champion cut 95 percent of the merchantable timber off its land by the late 1980s. Plum Creek plans to complete its Montana timber harvest program by 2000. The ecological effects will last for decades.

Champion decided to liquidate its timber in 1981 to raise money for the firm's expansion. It planned to build new mills in Texas and Michigan and needed capital. The most liquid capital it had was forests covering the mountains of the northern Rockies.

Plum Creek is the second-largest private timber owner in the Pacific Northwest behind Weyerhaeuser. This regional timber giant is the step-child of Burlington Northern, the railroad company that had vast land holdings after its formation from the merger of the old Northern Pacific and Burlington railroads. James J. Hill built the Northern Pacific Railroad from Lake Superior to Seattle and Portland on the financial roadbed laid by a 40-million-acre land grant from the federal government in 1864. Most of the land has been sold to farmers and timber companies, but the company kept 1.4 million acres of old-growth timber in Washington, Idaho and western Montana. In 1968, the company bought Plum Creek Timber Company, a small Montana lumber operation, and sent its loggers into the woods to begin cutting the railroad's vast holdings. In 1980, Plum Creek accelerated its harvest at the same time that Burlington Northern began restructuring its debt and its oil, timber and mineral assets to prevent a hostile takeover.

In the early 1980s, when stock prices were low, assets such as timber lands became burdens. Companies became worth more than their stock market value and were ripe for what became a common practice during the decade, corporate raiding. Stockbrokers used to say you could buy timber cheaper on Wall Street than on the national forests. The hostile takeover was a relatively simple proposition. Buy a company by going into deep debt, perhaps with junk bonds, and pay off the debt by liquidating the company's assets.

To prevent just such a takeover, Burlington Northern stepped ahead of the raiders and saddled Plum Creek with $325 million in long-term debt and then sold the subsidiary to a limited partnership. Burlington Northern remained in control but now must sell off its assets to pay back the debt. During the 1980s, Plum Creek was cutting its timber at a rate of about 550 million board feet a year. By its own estimate, its timber lands were growing about 210 million board feet of wood annually. In short, Plum Creek was cutting at a level twice what the forests could sustain by silvicultural standards alone. In 1988, Plum Creek cut 680 million board feet of timber off its land, three times the annual growth rate.

The effect in the northern Rockies was devastating. Plum Creek is the largest private landowner of grizzly bear habitat, and much of what was pristine old-growth forest in 1975 today is six-inch-high tree farms. Grizzly bears have few places to run or hide. Since the land grant gave the railroad every other section of land from the public domain instead of one large block, Plum Creek lands are checkerboarded throughout the national forests, a pattern that has cumulatively damaged wildlife habitats and watersheds in much of the Rockies. In Montana's Swan Valley, a key corridor for grizzlies between the Mission Mountains wilderness and the Bob Marshall Wilderness Area, road densities increased to three and four miles of road per square mile. The proposed federal Grizzly Bear Recovery Plan recommends road densities of no more than one mile per square mile. The result is the potential isolation of the grizzlies in the Mission Mountains from the larger population in Bob Marshall, which in turn is connected with contiguous populations deep into Canada.

These cumulative effects already have forced the Forest Service to reduce its own projections for timber harvest in Swan Valley. Some timber previously planned for sale in the 1990s won't be sold for as long as twenty years to allow cover for grizzly bears. Similar reductions are planned throughout national forests in Montana and Idaho, causing the timber industry to howl in frustration. In northwestern Montana, Lolo National Forest supervisor Orville Daniels sharply reduced the planned harvest of his forest partly because of the massive cutting on Plum Creek land. In many drainages, Daniels said, he can't meet water quality standards to prevent sedimentation and cut any more trees. When the water quality suffered, so did fish habitat. The Intermountain Forest Industry Association appealed Daniels' decision to the head of the Forest Service but was unsuccessful.

The practice of clearcutting forests has the greatest effect on wildlife habitat. Clearcuts open up interior forest areas to sunlight, increasing the temperature of the soil and changing the plant community and ultimately the animal community as well. With the overstory gone, the forest community changes all along the food chain. For grizzly bears and wolves there are mixed results. The loss of hiding areas and the introduction of increased human activity are the most harmful. Also, the roads bring human pressure to bear on wolves, usually in the form of hunters, poachers or angry ranchers seeking to eradicate them. Clearcuts also encourage so-called weed species, such as whitetail deer. At first glance, it would appear that encouraging whitetail production would benefit wolves, since deer are wolves' favorite food. But since the clearcuts are almost always associated with roads and human activity, the clearcuts actually serve to attract wolves into areas where they conflict with humans.

From 1980 to 1989, trees were harvested from 704,001 acres of national forest in Idaho and Montana, mostly along the northern Rockies. Clearcutting accounted for at least 334,073 acres, or roughly half the total acreage, Forest Service figures indicate.

As was shown earlier in this book in the cases of the northern spotted owl, the Bruneau hot springs snail and the grizzly bears of the

Cabinet-Yaak and Fishing Bridge, politics, not science, was the driving force for the 1980s timber harvests. In 1989, the supervisors of the national forests in Region 1, which includes Idaho and Montana, wrote to Forest Service chief forester Dale Robertson saying that the agency was out of control. They asked him to restore balance to their programs and budgets and not to favor commodity interests. While Robertson paid lip service to the agency's "New Perspectives Program," designed to take ecological values into account, he continued to pressure his regional foresters to meet what were clearly unrealistic timber harvest targets. John Mumma, Region 1 chief forester, said in 1990 that his professional success was tied to meeting those inflated targets. Mumma was 30 percent short that year. The next year he was pushed out the door, given the chance to retire or take reassignment in Washington, D.C. He chose retirement and testified before the House Subcommittee on Civil Service that he was forced out of his job because he didn't allow enough trees to be cut to satisfy three Republican congressmen.

"All I tried to do was perform my job as a career civil servant and to carry out the policies of the executive branch in accordance with federal law," Mumma said in tearful testimony. "I have failed to reach timber quotas only because to do so would have required me to violate federal law."

The pattern was all too familiar. U.S. senators Larry Craig (R-Idaho) and Conrad Burns (R-Montana) and former U.S. Representative Ron Marlenee (R-Montana), at the behest of the timber industry, had been pressuring Robertson to meet the timber harvest targets. These targets were set artificially high in the national forest management plans written in the early 1980s due mostly to the political pressure of the timber industry and its major political benefactor, Idaho Republican senator James McClure, Larry Craig's predecessor. Craig had learned the game well from McClure. In a letter of May 23, 1990, Craig told Dale Robertson: "You have a serious management problem that must be addressed. It is my hope that you will move to assure targets are met and line officers held accountable for targets." While Craig, Marlenee and Burns denied placing undue political pressure on Robertson to oust Mumma, the timing could hardly be considered coincidence.

The massive harvests on the American side of the border were more than matched by the Canadians. The two nations' timber-cutting programs threatened to have serious ramifications for the large endangered species of the northern Rockies. The survival of the mountain caribou and the gray wolf, especially in the northern Rockies, has depended up till now on Canada. The north-south orientation of the Rockies and other smaller mountain ranges tends to funnel the movements of animals and humans alike. As more and more humans entered the region from the south, the animals moved north. Once in Canada, they found a huge, largely untouched wilderness where they could thrive unbothered by civilization.

In the 1980s this pattern changed. British Columbia had embarked on a major program to harvest its forest resources. In 1986, some 4.1 billion board feet of timber was harvested in Oregon and Washington. By comparison, British Columbia cut 5.24 billion board feet, mostly from provincial lands. Up till the present day, those lands have been managed primarily for timber harvest, with even less protection for wildlife, recreation and other values than U.S. laws provide on national forest land. Entire mountainsides have been left bare, triggering mudslides and sedimentation of streams.

"The impact on our natural environment in south British Columbia is astounding. Thirty years ago one in ten drainages was severely impacted. Now it is eight in ten," says Ralph Moore, an environmentalist in Creston, British Columbia. "What we've done in the last ten years is more than what's been done in the last ninety."

Creston is in a valley that is a very warm fruit-growing area only a few miles north of the U.S. border and Idaho. Creston is in the heart of the Yaak Mountains, or, as they are known in Canada, the Purcells. Moore, a member of the East Kootenay Environmental Society, has a peach tree in his backyard. Only a hundred miles north, icefields and glaciers dominate the landscape, even in the summer. The Creston area's range of climates closely matches the range of climate from the north to the south in the northern Rockies. This diversity has made the area rich in wildlife, including bears, wolves, caribou, mountain goats and most of the other species that inhabit both dry and wet areas of the region. From ponderosa pine forests in the valleys to the groves of hemlock and spruce in the inland rain forests of the mountains, the area, like most of British Columbia, has been heavily harvested by Canada's timber industry.

"It's being logged to death," Moore says. "Another twenty years at the rate we're going and all of the merchantable trees will be cut. The only trees that will be left standing will be those protected by topography."

Unlike his environmental colleagues to the south, Moore has few environmental laws to use to stop the logging. There is no Endangered Species Act in Canada, no National Environmental Policy Act or National Forest Management Act—the tools that U.S. environmentalists have employed to curb timber cutting in American forests.

Canadians have had to carry their fight to the political arena, where they have had some success. There is a growing movement in British Columbia to end the long-term tree farm licenses issued to a few large corporations to harvest provincial lands. Small loggers, Indian tribes and local politicians have joined environmentalists in a coalition working to cancel the long-term licenses and replace them with licenses or contracts under the control of local communities instead of large sawmill or pulp mill owners.

Steve Herrero, a biologist at the University of Calgary and one of the world's top grizzly bear experts, has called on the Canadian government to set aside additional large preserves to ensure the long-term survival of large mammals in Canada. He looks south at the fragmentation that has taken place in the last century in the American West and worries that it might be the future for Canada's so-far undeveloped western backcountry. National parks cover 12,000 square miles in the northern Rockies of Canada and are home to about 850 grizzly bears, about 3 percent of Canada's entire grizzly population, Herrero says. Grizzly habitat outside the parks is being fragmented and destroyed at a rapid pace. Most of the destruction is irreversible, according to Herrero. The development already begun in the region could continue at the same frantic pace for years.

"If you're looking into the long term you need protected habitat," he says. "Canadians are intensive natural resource users and I don't see a change in that in the near future. Only the strictly protected areas may be around 300 years from now."

The same may be true about the northern Rockies in the United States. Only the time frame is different. It could be within this lifetime that all but the national parks and wilderness areas are made useless for grizzlies. Caribou may never again move farther than a few miles south of the Canadian border. Only the prolific wolf appears to have a healthy chance of surviving among humans in the long term. But even that prospect depends on the premise that people will stop shooting them.

Setting aside large, uninterrupted ecosystems throughout the northern Rockies is impossible today. Humans are not going to back away

212

from the frontiers of civilization. The only hope is to halt destructive activities and begin to restore, where possible, the web of life necessary for the survival of the natural ecosystems. If we can restore and protect viable populations of the large, wild mammals, then we know we have healthy ecosystems.

Just keeping intact marginally protected natural ecosystems will be an overwhelming task. Montana's Swan Valley is part of the grizzly bear recovery area and therefore is supposed to be protected as bear habitat. The area, which has suffered from ambitious road building and heavy logging, is an example of what Chris Servheen, FWS grizzly bear recovery coordinator, calls a "fracture zone," where a supposedly intact and large ecosystem—the Northern Continental Divide—is in danger of fragmenting into two smaller areas. Another fracture zone is the Middle Fork of the Flathead River on the west side of Glacier National Park, where logging and roads threaten to block off passage through the river valley between big chunks of wild land.

In addition, Herrero and others warn that development on the boundaries of these areas threatens the quality of the habitat within. In Canada's national parks, Herrero warns that the grizzly bear population could decline 10 to 25 percent over the next forty years because of habitat losses just outside the park boundaries.

Many conservation biologists share Mike Bader's view that none of the natural ecosystems are large enough to preserve grizzlies, wolves and caribou for centuries. Logging and general human development and encroachment threaten to leave Canada's wild lands as ecological islands in the future. So while coordination with Canadian officials is a necessary part of the preservation of the health of North American ecosystems, it can't be the only hope. Americans cannot place the future of their grizzly bears, wolves and caribou in the hands of Canadian officials. The health of the ecosystems in the United States must rely on management south of the border as well.

To preserve the integrity of the northern Rockies, biological bridges must be created and maintained between the ecological islands, corridors or linkage zones that permit relatively free movement throughout the region. Just as thousands of miles of roads have been built by humans to allow them to travel around, biological rights-of-way must be built that follow the natural paths worn into the landscape by hundreds of generations of caribou, grizzlies and wolves. For where these large species can travel, so can the other species of the ecosystems on which they survive. The effects of inbreeding can be reduced, too, as immigrants add to the gene pool.

The need to protect or restore in some manner the wildlife corridors between natural ecosystems has been embraced by most biologists in the northern Rockies region. The agreement among scientists ends, however, with how much area must be managed or protected and how strict the management must be to biologically link the ecosystems. The Grizzly Bear Recovery Plan released in July 1992 proposes studies of linkage zones between the various recovery zones. The zones don't have specific boundaries, which could attract the kind of opposition triggered by the proposal in 1976 to designate critical habitat for grizzlies. Instead, they are general areas where biologists can study effects and behavior and then make more specific management proposals in the future. "We want to be sure what we do in these areas doesn't preclude use by grizzly bears in the future," Chris Servheen said.

One of the smallest but most important wildlife corridors lies just twenty miles north of Missoula, Montana. There the Burlington Northern Railroad and Highway 93 cross over Evaro Pass, forming a short barrier between the Rattlesnake Wilderness Area to the east and the Ninemile drainage to the west. Despite the highway and the railroad, wolves, and perhaps grizzly bears as well, have used this three-mile stretch of forest that drops right down to the right-of-way to move from the Bob Marshall–Glacier National Park area south into the Ninemile Valley.

The road has periods of heavy traffic, especially in the summer, as Missoulians and others use it as the gateway to Glacier and Flathead Lake. Servheen says protecting the narrow strip of forest that lines the pass, as well as the immediate areas behind it, is essential to preserving the link between these two important areas. Ninemile itself has become the valley of choice for wolves moving south through Montana toward Idaho. But without Evaro Pass, there could be no Ninemile corridor. Servheen is working with county officials and the Forest Service to preserve this precarious link. He is not proposing new, strict federal guidelines, preferring instead to leave management to local zoning ordinances. "This calls for a cooperative effort, not a regulatory one," he says.

There is a fear that because local governments are subject to pressures from developers they can offer only short-term protection to areas. Moreover, the highway itself could become a barrier to wolves and grizzlies as traffic increases. The Flathead Valley to the north is one of the fastest-growing areas in the West, with Californians and others flocking to Kalispell and other communities near scenic Flathead Lake. This is bound to increase traffic on Highway 93 over the next twenty years and beyond. This is a problem that is predictable. It requires long-term planning and attention to the needs of bears, wolves and other creatures.

Another important linkage zone is the Centennial Mountains that run along the Montana-Idaho border west of West Yellowstone, Montana. This long, undeveloped mountain range, with the Centennial Valley and Red Rocks Wildlife Refuge to the north and Idaho's Sand Creek Desert to the south, is the critical link between central Idaho and the Greater Yellowstone Ecosystem. Bears and wolves need only cross Highway 20 at Targhee Pass in the Henry's Lake Mountains or the lush, wide-open Henry's Lake flats to move between Yellowstone and the Centennials. Once there, the animals can follow the Continental Divide west to the Beaverheads and cross either the virtually deserted Birch Creek Valley or the Lemhi Valley to the north into the Lemhi Mountains. From the Lemhis, bears and wolves have several paths to the Frank Church–River of No Return Wilderness or the Selway-Bitterroot Wilderness to the north in central Idaho.

Restoring and preserving the paths for bears and wolves in this area conflicts with the plans of others. The Forest Service has embarked on an ambitious campaign in the Targhee National Forest in Idaho to cut timber in the Centennials. This timber program threatens to make the mountain range useless to bears and other wildlife species, despite efforts by the Forest Service to preserve wildlife habitat. Moreover, the U.S. Department of Agriculture operates the Dubois Sheep Experimental Station on thousands of acres of the Centennials. In the past, biologists have been concerned that bears were secretly being killed by sheepherders on experimental station lands. Since these lands are managed almost exclusively for sheep grazing, an aggressive predator-control program is practiced, including the use of pesticides and poisons. Richard Knight, Yellowstone Interagency Grizzly Bear Study Team leader, says the intensive management has taken its toll on wildlife. "The birds don't sing on the sheep station," he quips. Since the station is a sacred cow of the still-powerful National Woolgrowers Association, simply closing it is not an immediate political option. If environmentalists can convince political leaders to move the station elsewhere, they could open a major corridor between Yellowstone and the rest of the northern Rockies and still protect the interests of sheep ranchers.

The Centennials may not be the only potential link between Yellowstone and the other ecosystems. Although nearly 200 miles separate the Greater Yellowstone Ecosystem from the Northern Continental Divide Ecosystem, biologist Harold Picton of Montana State University hypothesizes that grizzly bears may already be moving between the two areas on what he calls a "filter bridge," which allows intermittent travel, running north and south through the Helena, Deer Lodge and Beaverhead national forests. He cites increasing numbers of grizzly sightings in places such as the Tobaccoroot and Gravelly mountain ranges southwest of Helena, Montana, as evidence. The habitat of those mountain ranges could support grizzlies for periods long enough to enable bears to set up housekeeping and even reproduce, Picton believes. "The movements along the several hundred kilometers of the bridge are those of genetic flow through interchange between groups of animals rather than those of an individual animal moving the entire distance," he wrote in a 1990 paper.

Unfortunately, Picton's filter bridge was left out of the recovery plan's linkage zones. Chris Servheen says that even if bears are occasionally moving through the areas, there is too much private land along the corridor to manage as a linkage zone. Biologists are lining up on both sides of the debate, but in the meantime, development, timber harvest and other activities that make those areas less hospitable to bears continue.

The Alliance for the Wild Rockies has proposed the most aggressive plan for preserving wildlife corridors: the Northern Rockies Ecosystem Protection Act. This coalition of 135 environmental groups from Canada and the United States proposes designating more than 3 million acres of roadless lands lying between the recognized ecosystems as wilderness. Another 2.4 million acres would be set aside as corridor management areas. It also would designate nearly 11 million acres of wild land inside the ecosystems as wilderness, making the act the most far-reaching ecologically based protection plan in the Lower 48 states, providing federal protection to more than 15 million acres of roadless lands in the northern Rockies.

The act is the brainchild mostly of Mike Bader, a former seasonal ranger in Yellowstone National Park and now executive director of the Alliance for the Wild Rockies. Bader has spearheaded his group into backing the bill that national environmental leaders told him was unrealistic. Bader's logic was simple. Environmentalists had compromised so much on wilderness bills in Idaho, Wyoming and Montana for so long that even they couldn't tell political leaders what they needed and why. "Traditional wilderness bills," Bader wrote in 1991, "with boundaries defined by political compromise are obsolete. New laws must incorporate new information about ecosystem integrity and simultaneously satisfy the needs of many different sectors of society."

The roadblock to wilderness protection in Idaho and Montana in the 1980s was the two states' congressional delegations. Represented by Idaho Senator McClure, the timber industry, miners and ranchers could stop any bill that moved through the Senate. Since environmentalists carried similar clout in the House, with the initial support of Representative John Sieberling (D-Ohio) and later Representative Bruce Vento (D-Minnesota) as chairmen of the House Interior Subcommittee on Fish, Wildlife and Parks, wilderness legislation was at a stalemate.

Bader and the Alliance for the Rockies' first president, Cass Chinske, a former Missoula city councilman, proposed breaking the deadlock by "rolling" the state delegations. Instead of waiting for a bill that was acceptable to the conservative congressional delegations, Bader and Chinske proposed building a national constituency for northern Rockies wilderness that eliminated the need for local political support. Their models were the Alaska Lands Act of 1980 and the Tongass National Forest Reform Act of 1990, where Alaska's congressional delegations were simply overpowered by the lobbying of national environmental groups.

The bill, based on the scientific principles of preserving biodiversity, was drafted by the Alliance for the Wild Rockies and introduced in the House in September 1992 by former Representative Peter Kostmayer (D-Pennsylvania). It contained some innovations, including the formation of a national Wildlands Recovery Corps, based on the Civilian Conservation Corps of the 1930s. The corps would restore more than 500,000 acres of key wildlife and fish habitat destroyed by logging and other activities. Heavily roaded areas, such as Swan Valley, could be restored to their natural habitat productivity by hiring unemployed loggers and road builders to close the roads, recontour slopes and excavate sediment from spawning areas.

Bader's recovery corps may sound like a pipe dream, but it is modeled on actual legislation that has been approved and funded by Congress. The 1978 Redwood National Park Expansion Act added 48,000 acres of heavily roaded and logged lands to the California park. In addition, a plan for restoration of the area to improve salmon and steelhead runs and preserve remaining stands of ancient redwoods was included. From 1979 through 1987, more than 147 miles of logging roads were removed and covered with natural vegetation. Thousands of acres of watershed were rehabilitated and replanted. The work generated a payroll of more than $24 million as well as millions of dollars in contracting and equipment. Of course, all of this is taxpayer money, hard to come by in a time of huge budget deficits. Bader argues that restoration funds could replace the federal funding already sent to the region to subsidize timber sales and road building by the national forests.

Bader has enlisted support for the bill from a wide range of scientists, including John and Frank Craighead and Charles Jonkel, three of the premier grizzly biologists of the region. He also has found support among several rock stars, including singer-songwriter Carole King, Bob Weir of the Grateful Dead and John Oates of Hall and Oates. These unlikely allies march on Washington from time to time, pressing for the bill or to fight the passage of other wilderness bills that threaten to open for development the remaining 5 million acres of roadless land in Montana or the 9 million acres of roadless land in Idaho.

Bader's vision is powerful, but his no-compromise strategy and negotiating demeanor—somewhere between a grizzly bear just awakened from hibernation and a cornered wolverine—have not endeared him to natural allies in Washington. Even if the heart of the Northern Rockies Ecosystem Protection Act eventually passes, as many environmentalists believe could happen, it probably will be someone other than Bader leading the final charge.

The standard assumption that underlies opposition to the act and to any other wilderness protection plan in the northern Rockies is that it will cost the states jobs, especially in the timber industry. The industry itself argues that if the timber available in roadless areas in Montana

and Idaho is not available in the next two decades, then rural communities that now base much of their economic health on local sawmills
will shrivel up and die.

These gloom-and-doom predictions make great political rhetoric in
the two states, where the timber industry has defined the debate for
decades. In every single environmental battle since the 1950s the timber industry has argued that protecting wildlands would cost jobs in
the economies of the two states. The truth is that Idaho already has
more than 4 million acres of designated wilderness and has lost few
timber jobs due to wilderness designation. Montana has more than 2
million acres protected, and its timber economy has not shown employment fluctuations tied to their preservation.

Where wood-products jobs have dropped, it has been mill efficiency and modernization that have led to the decline. In Montana,
wood-products employment peaked in 1979 at 11,600 workers. In
1987, loggers harvested a record 1.376 million board feet of timber,
yet only 9093 people were employed in the industry. In Idaho, the
number of lumber mill jobs dropped from 18,800 in 1979 to 14,600
in 1989, a 22 percent loss at a time when timber harvest was steady
or increasing.

Thomas Power, chairman of the economics department at the University of Montana in Missoula, says flatly that protecting wildlands
not only won't hurt the economies of the two states but instead will
save them. He makes the case in two studies of the impacts of the
Northern Rockies Ecosystem Protection Act on employment in Idaho
and Montana. Power wrote: "Protected landscapes are a crucial part of
the economic base of Montana [and Idaho] and these high quality
natural environments have provided ongoing vitality in Montana's
[and Idaho's] local economies despite the ongoing decline in employment in extractive industries. Further damage to that landscape
through extension of roaded logging into Montana's [and Idaho's]
remaining wildlands threatens Montana's [and Idaho's] economic future while providing very few current jobs."

Power says if wilderness protection were to be extended to virtually
all of the remaining Forest Service roadless areas in both states, only
about 1000 actual forest-industry jobs would be lost and only 2200
total jobs would be lost. The 1600 jobs that would be lost in Idaho
represent only one-third of one percent of the approximately 500,000
jobs in Idaho. The 600 jobs that would be lost in Montana represent

two-tenths of one percent of the 380,000 jobs in Montana's counties that have national forests, Power says.

Since the Montana economy has been generating about 6000 addi-
tional jobs each year in mostly unrelated fields, the 600 jobs lost would be replaced in about five weeks. In Idaho, the economy has been generating about 11,000 new jobs each year, so it would take about seven weeks to replace the 1600 jobs lost in the timber industry by protecting wilderness.

"It needs to be emphasized that the ongoing expansion of the Idaho economy is at least partially tied to its high quality natural landscapes and recreational opportunities," Power wrote. "Degrading these damages Idaho's economic base and the economic well-being of its population. In that sense, these job losses tied to reduced timber harvests are not net job losses at all. They are more than offset by the ongoing expansion of the economy supported by these protected landscapes."

The myth that extractive industries are still the backbone of the economies in the two states dies hard. But the economic facts are hard to dispute. In a study of the economy of the three-state Greater Yellowstone Ecosystem, Ray Rasker, an economist with The Wilderness Society, showed that from 1969 to 1989 growth in employment and income from agriculture and the extractive industries in the twenty-county region was less significant to the economy than growth in other sectors. In 1969, the extractive industries employed one of every three workers in the twenty counties. By 1989, those industries accounted for one worker of every six in the region.

Power found a corresponding statewide trend in Montana. During the first half of the 1980s, wage and salary jobs shrank by tens of thousands while overall employment grew. The new jobs were self-employment jobs, entrepreneurial enterprises of people creating jobs for themselves and others.

Rasker reported that a wide range of companies—pharmaceutical firms, mail-order catalog companies, publishing companies and even specialized manufacturing companies—had moved into the Greater Yellowstone area during the 1980s. Advances in telecommunications technology during that decade and access of rural communities to fiber-optic phone lines, fax machines, computer links and satellite technology have spawned entirely new kinds of "knowledge-based industries that can locate anywhere in the world."

"People care where they live," Power said in a speech to the Greater

Yellowstone Coalition in 1990. "The quality of the natural and social environment is a dominant force in determining the location of the population." He cited the examples of the flight to the suburbs and the migrations to the Southwest desert and to the Sunbelt during the 1980s. In all these cases, Power said people moved away from both jobs and commercial centers to areas with lower wages. "People took significant risks and made significant sacrifices to obtain the living environments they wanted."

Other groups that can choose to live wherever they want and who are moving in increasing numbers to the northern Rockies are retired people. They take with them pensions, retirement plans and returns from past investments. Power says this represents 40 to 50 percent of the personal income flowing into the communities of the region, a flow of income far larger than that created by any single industry or combination of industries.

Some communities are actively recruiting retired people. Rexburg, Idaho, located thirty miles south of Yellowstone National Park, lost the corporate headquarters of Diet Center, a national weight-loss business, and with it 250 jobs. But the town of 14,000 found a new way to create economic activity at Rick's College, a junior college with an enrollment of 7000 students. The dorm rooms and apartments filled by students in the winter used to lie empty in the summer. But community planners recruited "sunbirds," retired residents who spend the winter in the Southwest, to spend the summer in Rexburg. Now, in a migration similar to the move from winter range to summer range by the area's elk and deer, more than 1500 sunbirds summer in Rexburg, lured by the beauty and natural wonders of the Greater Yellowstone Ecosystem.

Other communities, such as Jackson, Wyoming; Sandpoint, Idaho; and Kalispell, Montana, have lured entrepreneurs from California and other urban states who long for the recreational opportunities available in the wilderness areas near these communities. These communities also have flourished as tourism centers and jumping-off spots for hikers, hunters, fishermen and other backcountry visitors.

These new economic growth centers depend on the region retaining its natural qualities. The people of the northern Rockies cannot continue to chop down its forests and pollute its rivers and hope to preserve the region's natural character. Protecting the natural ecosystems that allow the survival of wolves, caribou and grizzly bears also preserves the awesome landscape and its subtle splendor. Some will argue that the region's physical attributes can be protected without the extraordinary measures needed to preserve its wildest animals. But scenic beauty alone doesn't make the Rockies wild. Without the grizzly bear, the gray wolf, elk, bighorn sheep, bald eagles and salmon, the region would be like a cardboard backdrop in a cheap photo studio, a caricature of its former untamed character.

221

The clash between the old vision of the northern Rockies, in which human prosperity depends on damaging, extractive industries, and the new vision, in which both humans and other species depend on preservation of natural life cycles, dominated the political debate in the 1980s. Yet people on both sides of the debate, people who live and work in the region, share many of the same values and emotional ties to the landscape. Dan Kemmis, the mayor of Missoula and author of the book *Community and the Politics of Place*, says people not only choose the place they want to live but also are chosen by that place. Just like the grizzly bear, bison, and cutthroat trout, humans are tied to the habitat that best meets their needs.

"We rarely make the connection between people and the land as we do naturally with ecosystems and other life forms," Kemmis says. "We all know that great bears are shaped by the place they inhabit. The same is true of people. People are chosen, are shaped, by the places they choose to live."

Natives and newcomers, ranchers and city-dwellers, loggers and environmentalists share common challenges and experiences in the northern Rockies. No matter what happens, many of the conditions of life in the region, as well as in the entire world, will change in the next century. If over the next fifty years we alter the basic elements that make the northern Rockies or the entire Pacific Northwest a place where humans, grizzly bears, caribou, wolves and salmon can live, then it will be not only the wild species that are endangered but also the human inhabitants of the region. Humans may remain, but they, like the landscape, will be tamed and subdued.

10. Saving All
the Parts

THE logic behind the federal Endangered Species Act is simple, says former U.S. Fish and Wildlife Service (FWS) director John Turner, and he goes back to his Wyoming heritage to explain it in simple terms.

"Growing up, I can remember going into the shop there at the ranch, and like any farmer-ranchers worth their salt we had a lot of stuff lying around, like old tractors," Turner says. "My granddad and my dad used to say, 'It's important to save all the parts. You never know where you're going to need them.' When you're not smart enough to know what parts you're going to need, why not save all the parts?"

Turner's observation, a folksy, western version of Aldo Leopold's own comments about the folly of discarding seemingly useless parts, was designed to reach a new audience concerning the importance of preserving biodiversity. Turner was thinking about his friends and neighbors on the ranches of the Greater Yellowstone Ecosystem and in the halls of the Wyoming State House, neither a bastion of progressive environmental thought.

Less than a decade ago, the terms *biodiversity* and *ecosystem* were viewed by many westerners with fear and apprehension. They were associated with big government and eastern environmentalists who wanted to control westerners' lives. The words were associated with plans to expand federal control over private lands outside Yellowstone National Park. Timber sales, grazing permits and mining projects that once were routinely approved suddenly were being halted to preserve biodiversity or because they damaged the integrity of ecosystems, interacting living and nonliving components of a loosely defined area. To ranchers and loggers, ecosystems were simply lines drawn on maps by environmentalists to stop them from living their lives the way they had for decades.

Today, protecting ecosystems and biodiversity still doesn't sit well with many of Turner's old colleagues, but the terms themselves have become part of the public's awareness. Still, the institutional tools to manage and protect ecosystems and biodiversity have not yet caught up with the public's understanding. Land is still managed along human boundaries, according to the delineations of nations, states, national forests and counties. In the early 1990s, the only comprehensive law available for preserving biodiversity is the federal Endangered Species Act of 1973. Despite its ambitious goals, the act has failed to "save all the parts" of the nation's ecosystems. Seven species have been declared extinct after they were placed on the endangered species list, FWS scientists say, and in the last ten years alone, 34 more species have become extinct while awaiting listing. Others estimate the species "death toll" to be as high as 300 since 1973.

With more than 3000 domestic species identified as candidates for protection under the act, many within the federal government and the environmental community are looking for a more efficient and perhaps less politically polarizing method of "saving all the parts," or preserving biodiversity. And their counterparts in the business community are looking for a system that allows them to develop resources with some reasonable guidelines and predictability.

A growing number of people on both sides of the debate believe that to protect biodiversity, resource managers and political leaders must take the same approach that intelligent Wall Street investors take: they must diversify their portfolios. They need to find methods of protecting as much of the ecosystem as possible, concentrating on those parts they already know are important or, from an investor's standpoint, reliable. Then they must hedge their bets against the unforeseen. Just as investors cannot know all of the factors that will affect the value of their investments, resource managers cannot predict with total accuracy what portions of their domain are necessary to the ecological stability of the whole area. When kokanee salmon levels took a serious dive in the 1980s in McDonald Lake in Glacier National Park, it affected eagle populations from northern Canada down to the southwestern United States. The extirpation of wolves from Yellowstone National Park has led to unnaturally high numbers of elk, with ripple effects throughout the ecosystem. Preserving flexibility is as important to land managers

224

in a time of ecologic uncertainty as it is for investors in times of economic turmoil. This doesn't mean locking up every acre in restrictive management. It does mean preventing activities that will cause irreversible effects.

Soon after taking the post of U.S. Environmental Protection Agency (EPA) administrator in 1989, William Reilly asked the EPA's Science Advisory Board to review a 1987 EPA study that had compared different environmental risks. That report, "Unfinished Business: A Comparative Assessment of Environmental Problems," looked at thirty-one environmental problems and ranked them in an order that generally placed threats to human health highest and ecological threats lower. The results of the study reflected the basic priorities in environmental protection that had guided the agency from its formation in the early 1970s, with protecting humans as its highest priority. The EPA Science Advisory Board, in its September 1990 report, took an entirely different approach. It recommended that the EPA "attach as much importance to reducing ecological risk as it does to reducing human health risk." The board's reasoning was simple. In the real world, there is little distinction between human health risk and ecological risk. "Over the long term, ecological degradation either directly or indirectly degrades human health and the economy," the board wrote.

"In short, human health and welfare ultimately rely upon the life support systems and natural resources provided by healthy ecosystems," it indicated. The board extolled the value of natural ecosystems, such as forests, wetlands and oceans, which it said were extraordinarily important for the health of the earth and humans. In its own ranking system, the board rated habitat destruction and species extinction as two of the highest risks to human welfare and natural ecology. In comparison, oil spills, groundwater pollution and radioactive materials were rated as relatively low risks. Other high-risk environmental problems included stratospheric ozone depletion and global climate change, both of which can have dramatic effects on biodiversity. It listed traditional pollution hazards regulated by the EPA, such as herbicides and pesticides, water pollution, acid rain and airborne toxics as medium ecological risks. Several of the same problems—air pollution, worker exposure to chemicals, and polluted drinking water—were ranked as higher human health risks. The board was calling on the EPA to stop regulating pollution piecemeal and instead to start managing ecosystems.

This new understanding of environmental risks has not yet spread throughout society. Still, the need to protect biodiversity and the natural ecosystems on which it depends has gained a wider audience in the last five years. In the Pacific Northwest especially, the battles over northern spotted owls, ancient forests, wilderness and salmon have forced everyone, from traditional resource managers to electric utility executives and aluminum industry leaders, to talk about protecting ecosystems.

The concept is simple. From the large animals at the top of the food chain down to plants and microbe communities, each species is important to the preservation of the whole. Even the genetic variations within each species are important to that species' long-term survival and the survival of the entire ecological community in which it lives.

The wild salmon in the Columbia and Snake rivers are a classic example. These fish have developed specific characteristics for adapting to their environment with the widest diversity possible. In the Columbia River Basin, more than 400 distinct stocks of salmon and steelhead once migrated up and down thousands of miles of tributaries. *Stocks*, a scientific term similar in definition to subspecies, is not necessarily recognized legally as such under the Endangered Species Act. Each stock has its own genetic blueprint—characteristics developed over thousands of years of evolution that make it uniquely suited to the tributary where it spawns, the rivers it migrates through and the section of ocean in which it matures.

Salmon cannot survive unless the three separate natural ecosystems in which they live remain intact and naturally connected. They need clean oceans with relatively stable currents. They need clean rivers and tributaries without barriers that prevent them from migration and spawning. They also need ecologically sound tertiary ecosystems surrounding the rivers and oceans so those ecosystems remain clean and stable.

Salmon are remarkably adaptive fish. Stocks leave the ocean at different times to start their migration back to their spawning grounds, providing insurance that if one seasonal event, such as a drought, prevents an entire stock's spawning, other stocks will still reproduce. Historically, spawning occurred nearly all year in the Columbia and Snake rivers. Some salmon spawn in tributaries near the ocean; others swim more than 900 miles inland before leaving their eggs and dying. They are programmed to return to the stretch of stream where they were hatched, but some fish stray to other places in the stream or even to other streams, ensuring the widest use of spawning habitat. These variations in behavior and needs not only helped protect the species but also transferred nutrients and energy from the rich Pacific to the more sterile high-elevation ecosystems. Historically, for example, the salmon runs provided the major food source for grizzly bears living in central Idaho and the Bitterroots. The lack of strong runs today is a major hindrance to bear recovery in those areas.

Much of the salmon's genetic diversity already is lost, making what remains even more valuable. Scientists estimate that there are about 200 separate stocks of salmon remaining in the Columbia River and its tributaries, only half of what was once there. The Snake River chinook and sockeye already have been listed as threatened and endangered, respectively. Many others are sure to follow, not only in the Columbia River Basin but all along the Pacific Coast. Eventually, without a major change in management, similar losses will take place in British Columbia, Alaska and even on the Russian shores of Siberia and Kamchatka.

Most attention has been focused on the Snake River salmon because of the significant potential for economic disorder from implementation of the Endangered Species Act. Despite their focus on the endangered stocks, officials have tried to take an approach that considers all of the salmon problems throughout its Pacific Northwest range—an ecosystem management approach. The actual management plans vary widely between industry and environmental groups, yet all are aimed at managing the entire salmon life cycle. Jack Robertson, Bonneville Power Administration (BPA) deputy administrator, even called its salmon proposal an "ecosystem management plan." Organizations such as the Columbia River Alliance, which represents Oregon and Washington irrigators, aluminum companies and utilities, say they want a balanced approach to salmon management that takes into account all of the threats to salmon survival throughout its life cycle. What they really mean is they want as few changes to the Columbia's hydroelectric system and dam operations as possible. Their approach, which is similar to that of the BPA, the Northwest Power Planning Council (NPPC), environmentalists and the National Marine Fisheries Service (NMFS), reflects a scientific basis barely considered a decade ago. Despite regular squabbles and outright battles among the various industry and governmental groups involved in salmon management, the debate has rarely followed the spotted owl model, pitting the species against those who want to protect jobs. Instead, each interest group has tried to find a solution that threatens its interest the least. Often this has led to some unusual finger-pointing. The utilities have called for strict limits on logging, mining and grazing throughout salmon range to limit the effects of sedimentation on spawning streams. Idaho irrigators want changes in Snake River dam operations in Oregon and Washington so they will not have to give up more of their precious water. During the next decade it will be up to scientists to decide what needs to be done on land, sea and in the rivers to restore and preserve the spectacular regional resources the salmon runs represent. Consequently, if the future of salmon is ensured, then the other natural values that contribute to the region's quality of life and economic health also will survive.

Unfortunately, the cost of restoration will be high, much higher than it would have been a decade ago. The same is true for the northern spotted owl and the old-growth forest ecosystem that is its home. It is less true for the large mammals of the northern Rockies and the natural ecosystems on which they depend, since most remain intact, albeit tenuously.

In each of these examples, the destruction of ecosystems and the declining numbers of species did not get serious attention until the federal Endangered Species Act was brought into play. Its critics argue that the act was meant to be wielded as a shield instead of a sword. Yet in its absence, those crying out in defense of the disappearing species and their ecosystems were defenseless. Use of the act as a sword or as any other contrivance has been an act of desperation on the part of environmentalists and indeed a region that could not force itself to take bitter medicine it knew was necessary.

In 1990, Representative James Scheuer (D-New York) introduced a bill in Congress designed to protect biological diversity. It would have required federal agencies to ensure that their actions are consistent with the goal of protecting biological diversity "to the maximum extent practicable." Some proponents say the bill could have had the far-reaching effects of the National Environmental Policy Act, which established the process for requiring environmental impact statements. So far, the proposal has picked up little support, leaving only the Endangered Species Act to protect biodiversity.

"The Endangered Species Act is emergency-room conservation biology," says Sara Vickerman, a Defenders of Wildlife regional representative in Portland, Oregon. "It's obvious that the fish and wildlife agencies aren't putting enough into anticipating problems."

Vickerman is one of a growing number of conservation biologists and environmentalists who are advocating a new approach to preserving biodiversity nationwide and even worldwide. They hope that by anticipating and reacting early to ecosystem threats they can prevent the polarizing battles that have marked efforts to protect other endangered species. By identifying sensitive species early and adjusting management programs when there is still a great deal of flexibility, the adjustments socially and economically will be easier.

Generally, preventive management remains an option well into the extinction process. Even after habitat has been fragmented, species can be preserved without the extraordinary efforts now needed to restore salmon and northern spotted owls. When population levels drop so low that the genetic diversity of the species, sex ratios and catastrophic effects threaten the entire species, officials are only managing a crisis.

One of the most promising approaches is a strategy known as gap analysis. Gap analysis is the process of identifying the individual species and plant communities that make up the biodiversity of a particular region or ecosystem and then mapping the existing preserves and protected areas. Once the information has been gathered, scientists can determine which species or communities are not protected.

At the University of Idaho, scientists have devised a particularly useful method for conducting gap analysis, which can become an early warning system for disappearing species. Using computers and a sophisticated data base known as the Geographic Information System (GIS), biologist J. Michael Scott and his team are able to graphically portray and manipulate such variables as species richness, distribution and land management and ownership.

By overlaying these kinds of information, Scott can identify gaps that occur in the land protection system and provide land managers and biologists with the information base they need to protect vegetation types and ecosystems.

In 7000 square miles of southeast Idaho, for example, Scott and his team identified twenty-three major vegetation types. Fourteen of the twenty-three were not included in any of the eleven preserves in the region.

These newly identified areas could include future endangered species. They could be protected today, without major costs, through wise
management. The advantage for commodity users of the land is that it   229
might prevent efforts to protect large areas of land when only smaller
areas are needed, says Gerry Wright, a National Park Service biologist
at the University of Idaho. "I think this procedure allows you to focus
your attention only on the pieces of land that will do the most good,"
Wright says.

For Scott, a research biologist with the FWS at the University of
Idaho, his work in gap analysis grew out of the frustration of running
the California condor recovery program. He and his colleagues spent
millions of dollars to protect the remaining condors in the wild but
finally had to give up and rely on a captive breeding program. "We
were in the emergency room of the Endangered Species Act and we lost
the patient," he says.

Then Scott went to Hawaii to study birds in the lush rain forests of
the fiftieth state. There he and other researchers mapped the ranges of
endangered species such as the Hawaiian hawk and soon learned that
the state's existing network of preserves was not adequate. Using
Scott's research, the FWS and the Nature Conservancy were able to fill
in some of the gaps by purchasing new habitat.

When Scott arrived in Idaho, he and other researchers refined the
process he had used in Hawaii. He brought the GIS mapping system
into play and combined it with maps of animal and plant distribution
and maps of protected areas. Scott's team looked for areas where many
species occur together. These areas of species richness, Scott says, are
the most important to protect biodiversity. By conserving the places
with the widest array of plant communities and the greatest variety of
animal species, states can protect their biodiversity with the least
amount of cost and economic disruption.

So far, Scott's strategy has received wide support. John Turner, the
FWS and Congress have embraced the strategy as a tool that could
revolutionize wildlife management. By the end of 1992, Oregon, California, Utah and Nevada had completed statewide gap analyses. Washington, Montana, Wyoming, Colorado, Arizona and New Mexico
already had begun working on their own gap analyses. Eventually,
conservation groups hope the program will be expanded to every state.

Once the areas of rich biodiversity are identified, the job of protection is easier but not necessarily easy. Many of the important areas in

Idaho that were identified by Scott were private land. In many cases the owners are not going to be interested in selling their land to the Nature Conservancy or the federal government. Some may be willing to protect the important wildlife habitat or plant communities out of their own sense of ethics or appreciation for nature, but dependence on the good intentions of landowners is only a short-term solution.

Whether protecting specific endangered species or biodiversity over a large area, the need to develop long-term, reliable protection on private land is both challenging and important. The Endangered Species Act can place limits on private property use. First, endangered species cannot be "taken," either directly by killing them or indirectly by forcing them out of critical habitat. Private property can be designated critical habitat, including privately owned water rights in the case of salmon or other fish. Such a designation limits the use of the property by the owner. In the case of water rights, it can have far-reaching effects, since most of the agriculture in the region is based on private ownership of water—that is, that private water users have priority over fish and wildlife. The most conservative interpreters of the Constitution's Fifth Amendment say that such limits, without compensation, represent the unconstitutional taking of private property. Just as removal of an endangered species from its habitat is considered a taking under the Endangered Species Act, preventing the use of property by landowners is considered under the law as a taking in certain instances. These two legal concepts are bound to clash. The U.S. Supreme Court has not ruled specifically on the private property ramifications of the act. Conservative legal groups, such as the Mountain States Legal Foundation, claim that the Endangered Species Act could violate the Fifth Amendment's prohibition of the taking of private property without compensation if it stops landowners from developing their land to the fullest. Environmentalists argue that the Constitution allows reasonable regulation to protect the public interest. In a 1992 case, *Lucas vs. North Carolina*, in which a shoreline protection ordinance prevented a landowner from building on his property, the court ruled that a government cannot limit all economic use of property without compensation. Unfortunately, the decision did not resolve how far regulation can go before it is considered a taking.

The controversy concerning management of endangered species on private lands has limited the inclusion of such lands in critical habitat. All private lands were left out of critical habitat for the northern

230

spotted owl. The private land issue partly prevented the FWS from designating critical habitat for the grizzly bear.

The concern about potential timber-cutting limits on private land   231
compelled many landowners in Oregon and Washington in the late 1980s to cut their timber before it was mature, out of fear that they would lose the right to cut if their land was declared critical spotted owl habitat. Free-market economists say such actions show the need for a new approach to protecting endangered species and biodiversity on both public and private lands.

John Baden, chairman of the Foundation for Research on Economics and the Environment, based in Bozeman, Montana, and Seattle, Washington, is one of a number of economists who have pressed for free-market alternatives for protecting environmental values. Baden, who teaches at the University of Washington Business School in Seattle, says that environmental management today is based neither on ecology nor economics. Instead, he says, it comes down to politics, and political compromises most likely will neither meet the needs of the environment nor the desires of private property owners and therefore will be unsuccessful. In the case of the Endangered Species Act and the northern spotted owl, Baden argues: "Narrow self-interests, most unlikely to include the owls, will rule the day."

Baden and forest economist Randal O'Toole, of Portland, Oregon, argue that instead of trying to force land managers and private property owners into preserving environmental values, environmentalists should use economic incentives to encourage stewardship.

O'Toole, thirty-nine, a pony-tailed, bearded forestry graduate from Oregon State University, is founder of the Cascade Holistic Economic Consultants and *Forest Watch* magazine. He has made a career of showing the economic folly of U.S. Forest Service plans that encourage building roads and cutting timber far above the return the agency gets from many of its timber sales. Just as economic incentives drive the behavior of private property owners, O'Toole says that budget incentives skewed toward higher timber harvests have compelled Forest Service managers to degrade wildlife habitat, overcut timber and erode watersheds. O'Toole advocates broader public land user fees so that hunters, fishermen, bird watchers, hikers and other public land users are equally contributing to the budgets of the Forest Service and Bureau of Land Management (BLM). O'Toole says more environmentally sound decisions will be made when land managers are afforded larger budgets to protect wildlife, recreation and wilderness.

O'Toole and Baden believe that changing the incentives for bureaucrats and private property owners will more likely preserve biodiversity and endangered species into the future. They both advocate what they call a "biodiversity trust fund," paid for with a percentage of the revenues generated from public land activities, including logging, mining, oil and gas development, grazing and recreation. All of these activities have some effect on wildlife habitat, Baden says, thus justifying the fees or taxes.

The proceeds could be used to purchase private lands or conservation easements on areas identified as important wildlife habitat, such as those identified using gap analysis. Or biodiversity fund "bounty payments" could be made to land managers or private landowners who successfully breed endangered species on their land.

The fund would be managed by a board of trustees that included economic and business leaders with conservation credentials, environmental activists and conservation biologists. Baden estimates that if public land resources were sold at market value and 10 percent of the proceeds were paid to the fund, it could generate $500 million to $1 billion a year for habitat protection. Compare that with the roughly $50 million a year now spent by the FWS for endangered species protection.

Instead of dreading the discovery of an endangered species on their land, O'Toole says, landowners will support the listing of species because it could mean new revenue sources. With budget-based incentives that encourage and reward species protection, land managers will avoid practices that degrade wildlife habitat on public lands because it will cost them money.

Although such an approach may seem too good to be true, the Defenders of Wildlife is attempting to use Baden's and O'Toole's principles to change the political landscape for wolves in Montana. The group will pay $5000 to any owner of private land in the northern Rockies who has wild wolves reproduce on his or her land. So instead of wolves being a liability to ranchers and other landowners in the region, they become an asset.

"Instead of being silent about the presence of wolves—or at worst taking the shoot, shovel and shut up approach—the landowner might notify wildlife officials to receive a reward," wrote Baden and Hank

Fischer, Defenders of Wildlife's northern Rockies regional director, in
the July 1992 issue of O'Toole's *Forest Watch* magazine. "Instead of
discouraging wolf habitation through various—possibly illegal—   233
means, the landowner might try to improve habitat. . . . A landowner
might even discourage his neighbors from killing 'his' wolves."

The incentive approach shows great promise in changing the way
people look at endangered species. But it also has its limits. The short-
term economic value of an endangered species might not outweigh the
value of a major real estate development, or a gold mine or some other
economic use of the land. Just as irrigated land is taken out of production in the Southwest for more valuable urban development use, some
other economic values could make endangered species protection less
profitable. Market incentives won't work by themselves; they must be
part of a long-term plan for protecting endangered species and endangered ecosystems.

Managing entire ecosystems instead of single species grew as a concept
during the 1980s. This concept has spread worldwide, but its roots lie
in the Pacific Northwest, specifically in the Greater Yellowstone Ecosystem.

In 1882, only ten years after Yellowstone National Park was established, General Philip Sheridan noted that the boundaries of the
world's first national preserve did not contain the ranges of its largest
game animals. Thus, more than a century ago, the idea of a "Greater
Yellowstone" was born. Sheridan suggested doubling the size of the
park, including much of the elk and deer winter range in Jackson Hole
and Montana in his proposal.

The modern concept of the Greater Yellowstone Ecosystem emerged
from the grizzly research of John and Frank Craighead in Yellowstone
National Park in the late 1950s. As they followed radio-collared bears
around Yellowstone, it became clear that the bears were not limiting
themselves to the park. As the two scientists analyzed the vegetation
and food base in the park, they recognized that the park simply didn't
include enough area to ensure the grizzly bear's survival. In the
Craigheads' early writings they called the range of the grizzly bear in
and around the park the "Yellowstone ecosystem." In 1977, when
Congress called them to testify about the designation of critical habitat
for grizzly bears, the Craigheads were talking about a Greater Yellowstone Ecosystem.

"We were talking about the Yellowstone ecosystem as early as 1959," John Craighead recalls. "I don't know if we just tacked 'greater' onto it. That just kind of evolved."

234

With all of the Craigheads' accomplishments in grizzly research, in their use of satellites for ecosystem mapping and in developing radio telemetry for wildlife research, perhaps their greatest legacy is that one phrase. Once it left their lips, it took on a life of its own.

Several environmental activists and national park officials took hold of the idea in the early 1980s. Rick Reese, in his book *Greater Yellowstone: The National Park and Adjacent Wildlands,* said he first heard the phrase from the late John Townsley, a former Yellowstone superintendent. Reese, then director of the Yellowstone Institute, and several environmentalists in the region, including Ralph Maughan of Pocatello, Idaho, a political science professor at Idaho State University, and Phil Hocker, an architect from Jackson, Wyoming, began talking about forming a regional organization dedicated to protecting Greater Yellowstone. Maughan recalled that the idea took three years to blossom before the Greater Yellowstone Coalition (GYC) was formed in 1983.

Meanwhile, the first federal Grizzly Bear Recovery Plan was released in 1982, and its recovery zones were called ecosystems. In the Yellowstone area, 9500 square miles in and around the park were called the Yellowstone Grizzly Ecosystem.

The management program put in place to recover the grizzly clearly called for cooperative management between the National Park Service and the Forest Service on a scale never attempted before. Greater Yellowstone was cut up like a jurisdictional pie between three states, two national parks, seven national forests, three Forest Service regions and several national wildlife refuges. Each was managed for different goals with different rules even among different national forests.

The new Greater Yellowstone Coalition was frustrated by the lack of a single program for managing the ecosystem. The Forest Service, National Park Service and other federal agencies each had their own goals and objectives for lands within the ecosystem. While the federal land managers practiced cursory coordination, there was no overall management goal for protecting the natural systems that crossed their boundaries. The GYC went to Congress for help in 1985. That year two subcommittees of the House Interior Committee held joint hearings on management of the Greater Yellowstone Ecosystem. The resulting analysis documented coordination problems among the various state and federal agencies managing the land in the ecosystem and recommended several changes. Faced with the possibility of congressional directives forcing more intrusive changes, the federal agencies, led by the Forest Service, decided to do their own study of coordination problems and seek ways of working together more efficiently.

The first result was a written compilation of the existing land management plans in the ecosystem. The agencies, especially the Forest Service, were afraid to use the word *ecosystem* in this first report confirming coordination problems. The word had become charged in the region's communities, which still saw themselves as logging, mining and ranching towns. The idea of a Greater Yellowstone Ecosystem represented to them a plot to expand the boundaries of Yellowstone. An earlier proposal of environmentalists had been to place a twenty-mile buffer around the park, but the idea had been quickly dropped. Unfortunately, the threat of such a proposal continues in the minds of ranchers and farmers even today.

The report by the Forest Service and National Park Service, released in 1988, delineated an area 31,000 square miles in and around Yellowstone and Grand Teton national parks as the "Greater Yellowstone Area." The report convinced land managers in the area that more had to be done to bring themselves together under a single "vision." A task force was formed, and the ambitious "Vision for the Future: A Framework for Coordination in the Greater Yellowstone Area" was drafted and the draft released for comment in 1990. In the report, even the Forest Service was unabashedly calling for ecosystem management in Greater Yellowstone, a 180-degree turn from its earlier reluctance to use the term *ecosystem*. The vision document's goals included conserving the sense of "naturalness" and maintaining ecosystem integrity, encouraging biologically and economically sustainable opportunities and improving management coordination.

Environmentalists had hoped that the vision document would set specific management parameters, such as the size of clearcuts, the number of roads allowed or the amount of grazing to be permitted within the ecosystem. Still, the document offered the most progressive, unified view the federal agencies had developed cooperatively on ecosystem management anywhere in the country. Unfortunately, its lack of specifics allowed critics from commodity user groups to use the vision document as a rallying point to rail against the growing consensus for ecosystem management.

236   Critics made the vision document appear as a sinister threat to the traditional timber, mining and ranching industries, not only in Yellowstone but throughout the West. A new group of players, calling themselves the "wise use" movement, used the vision document to spread fear in communities already hurt by declining timber sales and depressed livestock prices.

Dennis Winters, a former lobbyist for Saudi Arabia, had come to Montana to organize the fledgling wise-use movement, and the vision document was the perfect target for organizing against. Winters blamed environmentalists for everything from unemployment to increased wife beatings in small Montana communities. He told his ever-growing crowds in 1990 and 1991 that environmentalists were using the Endangered Species Act and the Yellowstone vision document to take control of their lives.

"These environmentalists are playing you like a cat and mouse. We don't have a resource problem. We have a political problem," he told the Greater Yellowstone Association of Conservation Districts at a 1992 meeting in Idaho Falls, Idaho. "The truth is so-called environmentalists don't want any of us here."

People for the West, an organization heavily funded by the mining industry and other commodity user groups, bussed hundreds of people to hearings in Montana, Idaho and Wyoming in January 1991 to protest the vision plan. Environmentalists, while not satisfied with the plan, found themselves on the defensive. The organizing of Winters and more traditional environmental adversaries, such as the Wyoming Heritage Foundation, which represents the oil, timber and mining industries, led to the seventy-two-page plan being cut to only eleven pages of vague goals. More than 7000 people commented on the vision document either at the eight public hearings or in writing, with most opposing it. It was a watershed event, since previously environmentalists had always marshaled their forces better than commodity user groups and had weighted public comments in their favor.

More important than the grassroots opposition were the efforts of the powerful special-interest groups—the timber industry, oil and gas lobbyists and others who found experienced voices in the Wyoming and Montana congressional delegations willing to bring pressure to bear on federal bureaucrats. The lawmakers took their opposition all the way to the White House, according to congressional testimony by Lorraine Mintzmyer, who was the Park Service's Rocky Mountain regional director in Denver. She testified in September 1991 before the House Subcommittee on Civil Service that Scott Sewell, the Department of the Interior's principal deputy assistant secretary for Fish, Wildlife and Parks, told her that White House chief of staff John Sununu told him that the vision document was a "political disaster" and had to be rewritten. Sewell told her that the Department of the Interior had delegated the rewriting job to him and that he was to make it appear as if it were the product of professional and scientific staff. Even though the final draft was a political whitewash, Sewell wanted it to look like a scientific document. He was told to cover his tracks.

The final eleven-page document was released in September 1991. Gone was the word *vision* in the title. Now it was simply a framework for coordination. Also gone were the lofty goals to preserve natural qualities and encourage sustainable economic opportunities and the guiding criteria for carrying them out. The goals were replaced with "principles," with no management recommendations.

Mintzmyer was transferred to the Park Service's eastern regional office, where she eventually retired early and filed suit against her old bosses. John Mumma, the Forest Service's Region 1 chief forester in Missoula, who also helped draft the vision document, retired after being given the choice of retirement or reassignment to Washington, D.C. Mumma already had gotten into trouble with the Bush administration for being unable to deliver what he said were unrealistically high timber harvests in the northern Rockies. He testified in the same hearing with Mintzmyer that politics, not science, was driving decisions in the Forest Service.

The flap over Mintzmyer's transfer and Mumma's departure brought to the surface the political meddling in resource decisions in the Pacific Northwest and the northern Rockies in the 1980s and early 1990s. From the Park Service's backtracking on closing Fishing Bridge, through the initial decision not to list the northern spotted owl, through McClure's manipulated consultations in the Cabinet-Yaaks, to the vision document flap, a consistent pattern of politics overriding science emerges. Yet through all of this, the move toward scientific ecosystem management continued.

For the Forest Service, the need to shift its management from primarily timber-sale planning to ecosystem management didn't start during the discussions about Greater Yellowstone. The national forest planning process that derived from the 1976 Forest Management Planning

Act had forced the agency to address a wide variety of interests in the 1980s that it had either ignored or only paid lip service to in the past. 238 The computer models it used to formulate the ten-year plans, known as Forplan, were basically designed to provide timber harvest estimates. Planners would feed into the computer what they called "constraints," such as wildlife habitat protection measures, water quality rules or old-growth requirements, and then see how those affected timber harvest. There could be no doubt that timber was the major product of the forest and all other values primarily were constraints to cutting more timber. Despite this obvious bias, the Forest Service, then top-heavy with foresters and road engineers, believed it was making balanced decisions.

Congress added to the pressures to emphasize timber management over protection of other values. Despite the computer plans, Congress actually set the timber harvest goals in the form of funding for timber sales and roads. When Congress appropriated timber-cutting and road-building funds higher than national forest requests, as was the case throughout the 1980s, the agency was pressed to find as much timber as possible. To do that, forest supervisors would ignore recommendations by their own biologists and hydrologists to protect wildlife and water quality. Even though the agency had an extensive public involvement process, national forests were managed by directives from Washington, D.C. This top-down process skewed the way decisions were made right down to the ranger district level. The foresters, who set up timber sales, and biologists, who were supposed to evaluate the effects of the sales on various wildlife species, often worked together only after the sales had already been set up. Instead of integrating the needs of endangered species and other wildlife right into a timber sale, the biologist would get involved later, perhaps making changes if wildlife problems were found. This process not only put biologists in an uncomfortable adversarial position, it was also less efficient, and this led to the large number of appeals from environmentalists, who would oppose the sales for the same reasons biologists would question them.

Foresters were trained in a discipline, dating back to the 1800s, that was based on the idea that silvicultural technology was a universally applicable science that could be applied to all forest sites to provide biologically sustainable forests. Only in the 1980s did foresters begin to seriously delineate between growing trees and growing forests. -

Generational factors helped drive the change in attitudes in the Forest Service. Those foresters who came into the Forest Service in the 1950s and 1960s were trained in the period when the agency was gearing up to fill a perceived national gap in private timber supplies. Passage of the Multiple Use–Sustained Yield Act of 1960 restated Forest Service founder Gifford Pinchot's call to manage the forests for a variety of uses; but even so, managing trees was still considered the main use of national forests. The difference in philosophies of a forester leaving college to work for the Forest Service and one going to work for the timber industry was negligible. After 1969, when environmental values began to creep into the curricula of forestry schools across the nation, a change of attitude began to emerge. A new generation of foresters was taught that a forest was not only a stand of economically useful trees as well as "decadent" or useless species, but an interconnected ecosystem.

The battles over the ancient forests of Oregon and Washington created the conditions to aid this widespread change in philosophy. Just as religious proselytizers find their most willing converts among the lonely and disillusioned, evangelists of the "new forestry" and ecosystem management found battle-weary foresters looking for salvation.

The high priests of this new, holistic forestry are predictably not foresters but ecologists. Oregon forest ecologist Chris Maser and Forest Service ecologist Jerry F. Franklin, a professor of forest resources management at the University of Washington Institute for Environmental Studies, have led the movement toward the new forestry. Maser and Franklin advocate new forest-management practices that encourage uneven-aged stands of trees and overall forest diversity.

Instead of clearcutting large sections of forest, the new forestry tells loggers to leave pieces of the natural forest intact, including large trees, snags, fallen logs and forest debris. Instead of one large, even-aged stand of a single species, this new approach results in smaller, mixed stands of multiple-aged trees.

While this appears to be a simple change, the science behind it is much more complex. The old forestry techniques called for cleaning up a harvest site and removing as much of the debris as possible. This removed an important source of nutrients, Franklin says, as well as fungi and other forest species that aid in tree growth and health. Coincidentally, other species that depend on the natural forest but are not necessarily important to foresters also benefit.

"The challenge to managers of public forest lands is to maintain ecological diversity in perpetuity," Maser and Franklin wrote in a joint paper published in 1988. "We must understand and accept biological diversity."

The new forestry inherently means cutting fewer trees, and that has made it a hard sell in the timber industry itself. Yet even companies such as Plum Creek Timber Company are experimenting with new techniques similar to those proposed by Maser and Franklin. Plum Creek is leaving buffers along streams, practicing select cutting and leaving more waste material on the forest floor.

Environmentalists also have been skeptical, worried that the new forestry simply offers a new way to rationalize logging in roadless areas and other previously unharvested areas in the region. Since most of the private land and roaded timber lands already have been cut in Montana, Idaho, eastern Washington and Oregon, most of the available timber is in areas previously off-limits to logging. Many of the areas have been proposed for federal wilderness designation, which would keep loggers out.

In the forests of eastern Oregon and western Idaho, where years of fire prevention and single-stand management have left millions of acres of highly valuable trees vulnerable to insects and fire, restoring forest health is the newest concept embraced by foresters. To those in the timber industry it means cutting as much timber as possible immediately to salvage dead and dying stands and restoring forest health. But to forest ecologists and a growing number of foresters in the Forest Service, it means restoring the forest to its more natural mix of species.

Maser cautioned those who would cut down all ancient forests and replace them with tree farms. Noting that tree-farm-type third-generation forests around the world are beginning to fail, he maintains that native forests must remain as models. "We did not design the forest, so we do not have a blueprint, parts catalog, or maintenance manual with which to understand and repair it," he wrote in the fall 1991 issue of *Inner Voice*.

Faced with managing the most important change in forestry philosophy since Pinchot was Dale Robertson, who became chief of the Forest Service in 1987. His own education and experience were firmly planted in the forestry philosophy of the past, yet he could not ignore the logic of the new forestry nor the rumblings within his own agency.

In 1989, Robertson met with all of his national forest supervisors, who told him in no uncertain terms that the agency was out of control, unbalanced toward commodity production and out of tune with its staff in the field. Robertson responded with a letter on February 23, 1990, telling all forest supervisors that standards and guidelines of forest plans should be followed, even if it meant compromising timber targets. The Forest Service began to embrace the new forestry in 1990, its centennial year, when it started its New Perspectives Program. The new initiative was designed to "sustain the greatest good" by "restoring, maintaining or enhancing the condition" of the national forests, Robertson wrote in a 1991 memo.

"This new direction reaffirms multiple use and sustained yield but with a difference," Robertson wrote. "Sustained yield is based more strongly on ecological principles to ensure that resource management sustains the overall health and productivity of the land." The initiative called for closer ties between national forest managers and scientists and set up demonstration projects nationwide.

There remained another side to Robertson and the Forest Service. Pressed hard by congressional leaders in Idaho and Montana, Robertson was forcing Mumma out of his post as Region 1 chief forester and pressuring others in the agency to meet unrealistic timber harvest targets, based on politics and faulty computer data. Mumma's departure came at the same time Robertson was calling for "new perspectives."

In June 1992, at the same time that world leaders were meeting in Rio de Janeiro at the Earth Summit, Robertson unveiled his plan to begin ecosystem management of national forests. "By ecosystem management," Robertson said, "we mean an ecological approach will be used to achieve the multiple use management of the national forests. It means that we must blend the needs of people and environmental values in such a way that the national forests represent diverse, healthy, productive and sustainable ecosystems."

While offering few specifics, Robertson told his supervisors to reduce clearcutting as a standard commercial timber harvest practice in the national forests. "In making future forest management decisions, clearcutting is to be used only where it is essential to meet specific forest plan objectives," he said.

Despite outward appearances of a full-blown ecological conversion, Robertson kept enough loopholes to ensure that clearcutting will continue on national forests for years to come. And he said that while the new approach may reduce timber harvests briefly, he was confident it would not cut timber harvest targets seriously over the long term.

Even though Robertson and the old guard in the Forest Service had to be brought kicking and screaming to this new approach, it was a dramatic change from the agency that mowed down much of the forests of the Pacific Northwest in the 1980s. Although ecosystem management won't free federal land managers from the influences of politics, it will, over time, increase the influence of their own biologists and ecologists. That can only mean more reliance on science than was seen during the destructive days of the 1980s.

In 1992, the Endangered Species Act came up for reauthorization. This means that Congress had to approve a base funding level for carrying out endangered species programs for the next five years. In each of the three previous reauthorizations, Congress used the reauthorization bill to make changes in the act. In 1978, the exemption process, which established the "God squad," was approved in the wake of the Tellico Dam decision. In 1982, the listing provisions were rewritten to ensure that only science could be considered in the decision to list. In 1988, minor changes were made to allow economic considerations to be more clearly recognized in recovery.

With the high-profile battles over the northern spotted owl and Pacific salmon, there has been a movement by Pacific Northwest Republican lawmakers, supported by a coalition of industry and wise-use groups, to weaken the act. Consequently, environmental groups have called for measures to strengthen the act.

The major debate over the act was delayed into 1993, and with other environmental issues, such as ancient forest protection in the Pacific Northwest, ahead of endangered species on the docket, most observers expect the reauthorization fight to stretch into 1994. In 1992, a coalition of industry groups, from the Farm Bureau to the American Forestry Alliance and the wise-use umbrella group Alliance for America, had proposed sweeping changes to the act that would allow economic considerations in the decision to list a species and would loosen land restrictions that could be enforced by the act. The Bush administration had been a major supporter of weakening the Endangered Species Act, and President Bush himself campaigned in the state of Washington criticizing the act's effects on rural timber-dependent communities.

House Speaker Tom Foley (D-Washington) has given some indication that he also would like to see the act changed.

Bill Clinton campaigned clearly and specifically in favor of pre-   243
serving the protections of the Endangered Species Act. He, too, fo-
cused his discussions on the issue of the Pacific Northwest and promised a forest summit early in his presidency to bring environ-mentalists, loggers, community leaders and scientists together in an attempt to heal the wounds that had been inflicted on the Pacific Northwest. Clinton won the election, carrying both Oregon and Washington, but losing in Idaho, Montana and Wyoming. That vic-tory replaced a strong opponent of the Endangered Species Act with a strong supporter, seriously undercutting the political hopes of the act's opponents.

Almost as significant was the reelection of Representative Gerry Studds (D-Massachusetts), who moved up to the chairmanship of the House Merchant Marine and Fisheries Committee. The committee has official jurisdiction over any legislation that involves the Endangered Species Act. Any amendment to the act must go through the Merchant Marine and Fisheries Committee and the House Natural Resources Committee. Studds, a stalwart supporter of the Endangered Species Act, faced a tough reelection fight in 1992 but over the years has been reelected easily. He, or the act's equally strong supporter, Representa-tive George Miller (D-California), chairman of the Natural Resources Committee, can be expected to short-circuit any attack on the act.

Support for the act is less certain in the Senate, where Senator Max Baucus (D-Montana) chairs the committee of jurisdiction, the Senate Environment and Public Works Committee. Baucus, who has a good environmental record overall, will face strong pressure from Montana groups unhappy with grizzly bear and gray wolf management. Still, any Senate bill must pass the House, and the best opponents can hope for is a compromise that will likely leave the strongest parts of the act intact. If anything, reauthorization this time around may expand the focus of the act from endangered species to endangered ecosystems.

Taking this larger ecosystem view makes sense and has benefits for both environmentalists and industry groups. Environmentalists would get a more efficient system for protecting the ecosystems that are the underlying focus of the Endangered Species Act in the first place. Industry groups would get a more efficient system that identifies the areas with the most sensitivity so that perhaps development can proceed without destroying the truly important pieces of landscape.

Clinton's Interior secretary, Bruce Babbitt, put this new concept to the test in the first major environmental initiative of the Clinton presidency. On March 25, 1993, Babbitt announced that he was protecting the California gnatcatcher, a four-inch bird with a kittenlike voice, as a threatened species under the Endangered Species Act.

Only 5000 of the birds remain in southern California, dependent on the rapidly disappearing coastal sage scrub for nesting. The coastal sage scrub ecosystem is not only the nesting grounds of the gnatcatcher, but also is home to fifty other species that are dwindling in numbers. The land also happens to be among the nation's most valuable real estate, and already 70 to 90 percent of the habitat has been destroyed by real estate development.

Environmentalists argued that the evidence showed the gnatcatcher was endangered, but developers said listing would threaten their constitutional rights to develop private property. By listing the species as threatened, Babbitt provided federal wildlife officials with enough flexibility to allow some of the gnatcatcher's habitat to be developed, even though it might cause the deaths of some gnatcatchers. Instead of focusing its attention just on the recovery of the gnatcatcher, Babbitt instructed the FWS and the state of California to develop an ecosystem approach to the coastal sage scrub habitat. Developers agreed not to develop some habitat to keep the coastal sage scrub ecosystem intact.

The deal, which Babbitt described as a "trailblazing effort," was risky for both sides. Environmentalists were forced to buy into a program that may not be enough to save the gnatcatcher or the ecosystem instead of using the uncompromising protection the act could provide. Developers gave up the opportunity to make millions on extremely valuable oceanfront land.

Babbitt's approach was in stark contrast to the listing battles of the Reagan and Bush years. Interior officials in the 1980s fought listing threatened and endangered species whenever there was economic controversy. They tried similar negotiated settlements in the absence of listing, such as with the Bruneau Hot Springs snail. But without the regulatory hammer of the act, developers had little incentive to keep up their side of the bargain, and environmentalists had nothing to ensure that any agreement would last.

Babbitt's new ecosystem approach, working within the powers of

the Endangered Species Act instead of around it, may lessen pressures
to gut the act when Congress finally gets around to voting on reauthorization. It also may give environmentalists the comfort level necessary
to make adjustments in the law to allow market incentives and other
experimental approaches to be tried to make it more efficient.

Clinton took the new approach into the promised forest summit,
renamed the Forest Conference, in Portland, Oregon, on April 2, 1993.
He faced an even greater challenge than Babbitt did with the gnatcatcher. Thousands of timber workers had already lost their jobs because of the northern spotted owl debate, and much of the most
important habitat already was lost. A decade of confrontation politics
spawned by the Reagan and Bush administration's polarizing rhetoric
had left anger and distrust between environmentalists and the timber
industry and its workers. But clearly, how successful Clinton is in
reconciling the environment and the ecology of the Pacific Northwest
will go far in demonstrating whether such a difficult task can be
accomplished nationally and worldwide.

The Endangered Species Act is strong enough already to protect the
remaining native ecosystems. Unfortunately, in most of the country,
development has destroyed most of the natural ecosystems that once
existed. In many others, development has advanced so far that restoring
natural ecosystems on a large scale is no longer possible. Outside the
Rocky Mountains, the Great Basin and the Pacific Northwest, perhaps
only the Sierra Nevada of California, the lake country of northern
Michigan, Wisconsin and Minnesota, the Adirondacks in New York
and the Florida Everglades could be managed effectively as natural
ecosystems today. Intact ecosystems remain in a few of those areas. In
many of them, significant restoration work would be necessary to bring
them back to a natural state. Northern Wisconsin and Michigan, for
example, once were covered with towering white pines. Today, the
more commercial red pine and aspen dominate the forest. The region
remains relatively isolated and even wild, yet many of the native species
are rare or gone forever. In Wisconsin and the rest of the United States,
conservation biologists are faced with the task of rebuilding biological communities. Years of human encroachment have torn apart the
natural systems and processes that make up a healthy ecosystem and
preserve biodiversity. For most areas it is already too late.

In the Pacific Northwest and northern Rockies, efforts to reintroduce wolves into Yellowstone and central Idaho and to restore

salmon runs are a part of the growing field of restoration biology. But unlike situations elsewhere, much of this huge area remains nearly as biologically intact as when pioneers settled there 150 years ago. It is the last place in the continental United States where ecosystem management or even bioregional management can be practiced on a large scale. It is there that Americans will decide whether to preserve other forms of life or continue the downward spiral that could eventually lead to human extinction.

The Pacific Northwest must be a model for bioregional, ecosystem management on a grand scale. If its inhabitants cannot preserve the special values of this wild place, then how can the rest of the nation or the world be expected to protect the wider biological wealth? It is there that Americans have "saved all the parts." The natural ecosystems that make the region such a special place to live in and visit won't remain intact if the welfare of its caribou, grizzly bears and salmon are only the consideration of environmental leaders or Washington policy-makers. People, shaping the land and shaped by it, are as much a part of the natural ecosystems today as are other creatures.

The Pacific Northwest won't be protected by outside forces. It must be the people in the region, who through good stewardship keep the region whole and sustainable. That means integrating the protection of natural ecosystems, both land and water, into all social, economic and political decision-making.

Protecting salmon must be part of any energy conservation or development decision made in the Pacific Northwest. Preserving the economic well-being of small forest communities must be linked with decisions to protect grizzly habitat. Stephen Kellert, a social scientist from Yale who specializes in the effects of human behavior on wildlife-management policy, said in a paper published in 1986, "To regard any economic system as environmentally separate, independent and superior is, in other words, to invite species degradation and decline."

The survival of grizzly bears in the northern Rockies may well be tied to the survival of small loggers or even the ranchers who pushed them to the brink of extinction. The survival of salmon may depend on the success of Pacific Northwest ratepayers in practicing energy conservation that reduces the need for electricity.

For the past decade, the political leaders and powerful special-interest groups that have controlled the allocation of resources in the Pacific Northwest and the nation have framed the debate in polarizing terms of jobs versus owls, or ecology versus economy. Even those who agree that ecology and economy are integrally tied are blinded by their biases. Some environmentalists see only the environmental constraints necessary, and some business groups see only the economic effects of those constraints.

Grizzly bears choose to live in a certain area because it meets their nutritional and territorial needs. If food is not available or if cover is too sparse, the grizzly bear may visit an area but won't stay. If humans build too many roads into a grizzly's range, the bear will leave. Humans choose their surroundings in much the same way, and in a sense, the country chooses them as well, says Missoula, Montana, mayor Dan Kemmis.

"If we looked around us at the other people who have been selected by the country," Kemmis says, "and recognize that even though we would still find many differences separating us, we would also begin to discover a greater capacity to solve the common problems that inhabiting this particular piece of country creates for us."

Those of us who are chosen by the Pacific Northwest and the northern Rockies don't mind winter and can enjoy all four seasons. We all like mountains, if only to stare at. The list of common characteristics are endless. Unfortunately, through most of the debates concerning natural resources, Pacific Northwest citizens seem to get caught up in their differences. We as a region, a nation and a world must find within ourselves the power to work through the differences in our values, needs and attitudes to build a common future.

Bibliography

Chapter 1   Endangered Species, Embroiled Region

Barker, Rocky. "America's Hinterland: Endangered Species Act May Change Lifestyles in Northwest." *Idaho Falls Post Register*, December 23, 1990.

Berry, Thomas. *The Dream of the Earth*. San Francisco: Sierra Club Books, 1988.

Bonneville Power Administration. "The Transmission System." In *What BPA Is All About*. Portland: Bonneville Power Administration, March 1985.

Brockman, C. Frank. *Trees in North America*. Racine, Wis.: Western Publishing Company, 1968.

Corn, M. Lynne, and Baldwin, Pamela. "Endangered Species Act: The Listing and Exemption Processes." CRS Report for Congress. Washington, D.C.: Congressional Research Service, Library of Congress, May 8, 1990.

Egan, Timothy. *The Good Rain*. New York: Knopf, 1990.

Haig-Brown, Roderick. Quoted on p. 5 of the May 1991 issue of *Transitions*, a publication of the Inland Empire Public Lands Council, Spokane, Wash.

Herndon, Grace. *Cut and Run: Saying Goodbye to the Last Great Forests in the West*. Telluride, Colo.: Western Eye Press, 1991.

Kadera, Jim. "Like Giant Fir Trees, Weyerhaeuser Co. Started Out Small." *Oregonian*, December 29, 1991.

Laverty, Kent J. *The Greening of Idaho's Economy*. Boise: Idaho Conservation League, May 1990.

Leopold, Aldo. *A Sand County Almanac*. San Francisco/New York: Sierra Club/Ballantine Books, 1974.

"Non-Agricultural Employment Growth in the Pacific Northwest, August 1990 to August 1991." In *Northwest Portrait 1992*. Seattle: Northwest Policy Center, University of Washington/U.S. Bank, 1992.

Patterson, J. H. *North America: A Geography of Canada and the United States.* New York: Oxford University Press, 1970.

250   Power, Thomas M. "The Employment Impact of the Northern Rockies Ecosystem Protection Act in Montana." *The Networker,* Spring 1992.

Rubovits, Robert. "Weyerhaeuser: A Report on the Company's Environmental Policies and Practices." New York: Council on Economic Priorities, May 1992.

Rudzitis, Gundars. "Progress Report: Migration, Sense of Place, and Non-metropolitan Vitality." *Urban Geography* 12(1) (1991): 80–88.

Rudzitis, Gundars. "Migration, Sense of Place, and the American West." Draft paper. September 1992.

Rudzitis, Gundars, and Johansen, Harley E. "How Important Is Wilderness? Results from a United States Survey." *Environmental Management* 15(2): 227–33.

Rutzitis, Gundars, and Johansen, Harley E. "Migration into Western Wilderness Counties: Causes and Consequences." *Western Wildlands,* Spring 1989.

Schwantes, Carlos A. *In Mountain Shadows: A History of Idaho.* Lincoln: University of Nebraska Press, 1991.

Schwantes, Carlos A. *The Pacific Northwest: An Interpretive History.* Lincoln: University of Nebraska Press, 1989.

Stuebner, Steve. "Can a New Plan Save the Fish?" *High Country News.* March 9, 1992, 1.

Turner, John, director, U.S. Fish and Wildlife Service. "Memo to all Employees. Re: Draft Strategic Plan." September 6, 1990.

Twining, Charles E. *Phil Weyerhaeuser, Lumberman.* Seattle: University of Washington Press, 1986.

U.S. General Accounting Office. "Endangered Species Act: Types and Number of Implementing Actions. Briefing Report to the Chairman, Committee on Science, Space, and Technology, House of Representatives. Washington, D.C.: U.S. Government Printing Office, May 1992.

Wilson, Edward O. "Threats to Biodiversity." *Scientific American,* September 1989, 108–16.

Chapter 2   Living with Grizzlies

Barker, Rocky. "The New Grizzly" (four-part series). *Idaho Falls Post Register,* September 13–16, 1987.

Barker, Roland R. "Trading Pork for Sheep" (editorial). *Idaho Falls Post Register,* October 4, 1989.

Charlier, Marj. "Counting Big Bears: Grizzlies Find Ways to Avoid the Government's Census in the Montana Rockies." *Wall Street Journal,* May 2, 1990, A8.

Corn, M. Lynne, and Gorte, Ross. *Greater Yellowstone Ecosystem: An Analysis of Data Submitted by Federal and State Agencies.* Prepared by the Congressional Research Service, Library of Congress, for the Subcommittee on Public Lands and the Subcommittee on National Parks and Recreation of the Committee on Interior and Insular Affairs. U.S. House of Representatives, 99th Cong., 2d sess., 1987.

French, Marilyn, and French, Steve. "The Baby Boom of 1990." *Yellowstone Grizzly Journal*, Winter/Spring 1991.

French, Marilyn, and French, Steve. "Yellowstone Bear Behavior During and After the Fires: A Report on the Foundation's 1988 Research Findings." *Yellowstone Grizzly Journal*, Spring 1989.

Gilbert, Bil. "The Great Grizzly Controversy." *Audubon*, January 1976.

Greater Yellowstone Coalition. *An Environmental Profile of the Greater Yellowstone Ecosystem*. Bozeman, Mont.: Greater Yellowstone Coalition, 1991.

Hammer, Keith J. "Interagency Grizzly Bear Committee Biased and Manipulative in Grizzly Bear Delisting and Monitoring Scheme." Unpublished report. Bigfoot, Mont., November 25, 1991.

Knight, R. R., and Eberhardt, L. L. "Projected Future Abundance of the Yellowstone Grizzly Bear." *Journal of Wildlife Management* 48(4) (1984): 1435–38.

Knight, Richard R.; Blanchard, Bonnie M., and Mattson, David J. "Yellowstone Grizzly Bear Investigations: Annual Report of the Interagency Study Team—1986." Bozeman, Mont.: National Park Service, U.S. Forest Service, Montana Fish and Game Department, U.S. Fish and Wildlife Service, Idaho Department of Fish and Game, and Wyoming Game and Fish Department, 1986.

Mattson, David J., and Reid, Matthew. "Conservation of the Yellowstone Grizzly Bear." *Conservation Biology*, September 1991.

McCormick, Bruce. "I Just Knew She Was Going to Kill Me." *Cody Enterprise*, September 17, 1990.

Reinhart, Daniel P., and Mattson, David J. "Bear Use of Cutthroat Trout Spawning Streams in Yellowstone National Park." *Bears: Their Biology and Management* 8: 343–50.

Richert, Kevin. "Grizzly Vs. Rancher: Mr. Egbert has had his run-ins with Mr. Bruin . . ." *Idaho Falls Post Register*, December 28, 1990.

Servheen, Christopher. "Grizzly Bear Recovery Plan." Missoula, Mont.: U.S. Fish and Wildlife Service, June 1992.

Shaffer, Mark L.; Samson, Fred B.; Perez-Trejo, Francisco; Salwasser, Hal; and Ruggiero, Leonard F. "On Determining and Managing Minimum Population Size." *Wildlife Society Bulletin*, no. 13, 1985.

Chapter 3   Troubled Waters

Barker, Rocky. "Columbia Gorge Loggers Feel the Effects of Owl's Federal Listing." *Idaho Falls Post Register*, December 23, 1990.

Englert, Stuart, and Barker, Rocky. "From Idaho to the Pacific Ocean." *Idaho Falls Post Register*, December 24, 1990.

Palmer, Tim. *The Snake River: Window to the West*. Washington, D.C.: Island Press, 1991.

Wimborne, Margaret. "Stanley Rancher Says Dams Blocked Salmon." *Idaho Falls Post Register*, December 23, 1990.

Chapter 4   Salmon Sacrifice

Anadromous Fish Law Memo. Issue 36, July 1986; Issue 38, November 1986;

Issue 40, March 1987; Issue 50, August 1990. Portland, Oreg.: National Resources Law Institute, Lewis and Clark Law School.

252 Barker, Rocky. "Biologist Spends Lifetime Fighting for Survival of Pacific Salmon." *Idaho Falls Post Register*, December 23, 1990.

Beiningen, Kirk T. "Fish Runs from Investigative Reports of Columbia River Fisheries Project." Portland: Pacific Northwest Regional Commission, July 1976.

Bonneville Power Administration. *Multi-Purpose Dams of the Pacific Northwest*. Bonneville Power Administration. Portland: Bonneville Power Administration, undated.

Broderick, Susan. Petition by Shoshone-Bannock Tribes of the Fort Hall Reservation in Idaho. Re: Listing Snake River Race Sockeye Salmon as Endangered. March 29, 1990.

Buck, Eugene H.; Abel, Amy; Kessler, Marla; and Bazan, Elizabeth B. "Pacific Salmon and Steelhead: Potential Impacts of Endangered Species Act Listings." CRS Report for Congress. Congressional Research Service, Library of Congress, November 16, 1990.

Chaney, Ed, and Perry, L. Edward. "Columbia Basin Salmon and Steelhead." Portland: Pacific Northwest Regional Commission, 1976.

Chapman, D. W.; Platts, W. S.; Park, D.; and Hill, M. "Status of Snake River Chinook Salmon." Don Chapman Consultants' report prepared for Pacific Northwest Utilities Conference Committee, February 19, 1991.

Chapman, D. W.; Platts, W. S.; Park, D.; and Hill, M. "Status of Snake River Sockeye Salmon." Don Chapman Consultants' report prepared for Pacific Northwest Utilities Conference Committee, June 6, 1990.

Cobb, John H. *Pacific Salmon Fisheries Appendix III to the Report of the U.S. Commissioner of Fisheries*. Washington, D.C.: U.S. Government Printing Office, 1930, 94.

Cohn, Lisa, and Henjum, Scott. "Salmon Savior." *Northwest Magazine*, April 1991.

Egan, Timothy. "It's Do or Die for Columbia River Salmon." *Spokesman Review*, April 1, 1991.

Evermann, Barton Warren. "A Report Upon Salmon Investigations in the Headwaters of the Columbia River, in the State of Idaho, in 1895, Together with Notes Upon the Fishes Observed in that State in 1894 and 1895." *Bulletin of the United States Fish Commission*, 1895.

Evermann, Barton Warren, and Meek, Seth Eugene. "A Report Upon Salmon Investigations in the Columbia River Basin and Elsewhere on the Pacific Coast in 1896." *Bulletin of the United States Fish Commission*, 1896.

Fish Commission of Oregon and Washington Department of Fisheries. Status Report: Columbia River Fish Runs and Commercial Fisheries, 1938–70. 1974 Addendum. Joint Investigational Report. Vol. 1, no. 5, January 1975.

Ford, Pat. "Does Salmon Barging Work?" *Idaho Conservation League News*, April 1992.

Gallagher, Don. "Dam Builders Tried Salmon Saving Theories Only to See Them Fail." *Idaho Falls Post Register*, April 10, 1992.

Nehlsen, Willa; Williams, Jack E.; and Lichatowich, James A. "Pacific Salmon at the Crossroads: Stocks at Risk from California, Oregon, Idaho and Washington." Portland: American Fisheries Society, September 28, 1990.

Northwest Power Planning Council staff. "Compilation of Information on Salmon and Steelhead Losses in the Columbia River Basin." Portland: Northwest Power Planning Council, 1988.

Oregon Department of Fish and Wildlife and the National Marine Fisheries Service. "Past and Present Abundance of Snake River Sockeye, Snake River Chinook, and Lower Columbia River Coho Salmon." Report prepared at the request of U.S. Senator Mark Hatfield. June 1, 1990.

Petersen, Keith C. "Idaho's Embattled Snake River Salmon." *Idaho*, Spring 1991.

Raymond, Howard L. "Effects of Dams and Impoundments on Migrations of Juvenile Chinook Salmon and Steelhead from the Snake River, 1966 to 1975." *Transactions of the American Fisheries Society* 108 (1979): 505–29.

Reed, Mary E. *A History of the North Pacific Division.* Portland: U.S. Army Corps of Engineers, North Pacific Division, 1991.

Sedell, James R., and Everest, Fred H. "Historic Changes in Pool Habitat for Columbia River Basin Salmon Under Study for Test Listing." Portland: USDA, Forest Service, December 17, 1990.

Titone, Julie. "The Ride of Their Lives: Barges Provide Fish Safe Passage." *Spokesman Review*, May 17, 1992, A1, A12.

Williams, Chuck. "The First Salmon." *High Country News*, April 22, 1991, 8.

Williams, Marla. "Fishermen Say They're Victims, Too." *Yakima Herald-Republic*, May 8, 1991, 6A.

Williams, Marla, and Simon, Jim. "The Struggle over Salmon." *Yakima Herald-Republic*, May 5, 1991, 1A.

Chapter 5   Lost Opportunity

Anderson, H. Michael, and Olson, Jeffrey T. *Federal Forests and the Economic Base of the Pacific Northwest: A Study of Regional Transitions.* Washington, D.C.: The Wilderness Society, September 1991.

Barker, Rocky. "Species Listing Scams." *Idaho Falls Post Register*, December 26, 1992.

Corn, M. Lynne. "Spotted Owls and Old Growth Forests." CRS Issue Brief. Washington, D.C.: Congressional Research Service, Library of Congress, October 12, 1990.

Chasan, Daniel Jack. "Whose Ancient Forest?" *Defenders*, September/October 1990, 16–38.

Durbin, Kathie, and Koberstein, Paul. "Lives in Transition." Special Report. *Oregonian*, November 29, 1990, 1–28.

Enos, Amos. "The National Marine Fisheries Service Endangered Species Program." Washington, D.C.: National Fish and Wildlife Foundation, 1990.

"Marbled Murrelet: Listing Held Up By Interior." U.S. Fish and Wildlife Service press release. *Greenwire*, July 29, 1992.

Taylor, Rob. "Whooo Cares?" *Government Executive*, April 1992.

Thomas, Jack Ward. "Wildlife in Old-growth Forests." *Forest Watch*, January/February 1992.

254   U.S. Department of the Interior. "Recovery Plan for the Northern Spotted Owl." Draft. April 1992.

U.S. General Accounting Office. "Endangered Species: Spotted Owl Petition Evaluation Beset by Problems." Report to the Chairman, Subcommittee on Fisheries and Wildlife Conservation and the Environment, Committee on Merchant Marine and Fisheries, House of Representatives. Washington, D.C.: U.S. General Accounting Office, February 1989.

Wuerthner, George. "Dammed River, Doomed Mollusks?" *Defenders*, May/June 1992, 12–13.

Chapter 6   The Sword and the Shield

Barker, Rocky, and Richert, Kevin. "Endangered Species Act Versus Future Development." *Idaho Falls Post Register*, December 27, 1990.

Bean, Michael J. "The 1973 Endangered Species Act: Looking Back Over the First 15 Years." *Endangered Species Update*. Ann Arbor: University of Michigan School of Natural Resources, August 1988.

Bouck, Gerald R., biologist with the Bonneville Power Administration. Letter of April 2, 1991, to Merritt E. Tuttle of the National Marine Fisheries Service.

Chase, Alston. *Playing God in Yellowstone*. Boston: Atlantic Monthly Press, 1986.

Devlin, Sherry. "Ten Years Too Late." *Missoulian*, June 23, 1990.

Durbin, Kathie. "Spotted Owl Loses Out in Lujan's Politicking." *Oregonian*, May 17, 1992, 1.

Fitzgerald, John M. "Withering Wildlife: Whither the Endangered Species Act? A Review of Amendments to the Act." *Endangered Species Update*. Ann Arbor: University of Michigan School of Natural Resources, August 1988.

Fox, William W., Jr. Biological Opinion Prepared by the National Marine Fisheries Service on the 1992 Operations of the Federal Columbia River Power Systems. April 10, 1992.

Greenwalt, Lynn A. "Reflections on the Power and Potential of the Endangered Species Act." *Endangered Species Update*. Ann Arbor: University of Michigan School of Natural Resources, August 1988.

Kasworm, Wayne F., and Their, Timothy J. "Cabinet-Yaak Ecosystem Grizzly Bear and Black Bear Research, 1989 Progress Report." Missoula: University of Montana, June 1990.

Reinhart, Daniel P., and Mattson, David J. "Bear Use of Cutthroat Trout Spawning Streams in Yellowstone National Park." *Bears: Their Biology and Management* 8: 343–50.

St. Clair, Jeffrey. "Politics Derails Biology: Marbled Murrelet Listing Deferred." *Forest Watch*, August 1992.

U.S. Department of the Interior, National Park Service. "Draft Environmental Impact Statement: Development Concept Plan—Fishing Bridge Developed Area, Yellowstone National Park." Yellowstone National Park, Wyo.: National Park Service, 1987.

U.S. Department of the Interior, National Park Service. "Development Concept Plan/Environmental Assessment. Lake–Bridge Bay, Yellowstone National Park." Yellowstone National Park, Wyo.: National Park Service, 1992.

Yaffee, Steven L. "Endangered Species Protection Through Interagency Consultation." *Endangered Species Update*. Ann Arbor: University of Michigan School of Natural Resources, August 1988.

Chapter 7   Roads to Recovery

Barker, Rocky, and Richert, Kevin. "Success Versus Failure." *Idaho Falls Post Register*, December 28, 1990.

Burnham, William A., and Cade, Tom J. "Peregrine Falcon Recovery Program: Status and Recommendations." Boise, Idaho: Peregine Fund, January 1992.

Clark, Tim W., and Harvey, Ann H. "Implementing Endangered Species Recovery Policy: Learning As We Go?" *Endangered Species Update*. Ann Arbor: University of Michigan School of Natural Resources, August 1988.

Corn, M. Lynne, and Baldwin, Pamela. "Endangered Species Act: The Listing and Exemption Processes." CRS Report for Congress. Washington, D.C.: Congressional Research Service, Library of Congress, May 8, 1990.

Mann, Charles C., and Plummer, Mark L. "The Butterfly Problem." *Atlantic Monthly*, January 1992, 47–70.

McClelland, B. Riley; Young, Leonard S.; Shea, David S.; McClelland, Patricia C.; Allen, Harriet L.; and Spettigue, Elizabeth B. "The Bald Eagle Concentration in Glacier National Park, Montana: An International Perspective for Management." In *Biology and Management of Bald Eagles and Osprey*, edited by David M. Bird. Ste. Anne de Bellevue, Quebec: Harpell Press, 1983.

Moser, Don. "One More Chance at Survival for the Cloud Runners." *Smithsonian*, April 1990.

Nalder, Eric. "Trials of Taxol." *Seattle Times*, December 15–17, 1991.

Peregrine Fund—World Center for Birds of Prey. *1992 Annual Report*. Boise, Idaho: Peregrine Fund, 1992.

Richert, Kevin, and Cecil, Molly O'Leary. "Recovery Takes Money." *Idaho Falls Post Register*, December 26, 1990.

Robbins, Jim. "Autumn of the Eagle." *National Wildlife*, October/November 1988.

Servheen, Chris. "Draft No. 1, Conservation Strategy for the Grizzly Bear— Northern Continental Divide Ecosystem." Missoula, Mont.: U.S. Fish and Wildlife Service, 1990.

U.S. Department of the Interior, Office of Inspector General. *Audit Report— The Endangered Species Program—U.S. Fish and Wildlife Service*. Report No. 90–98. Washington, D.C.: U.S. Department of the Interior, September 1990.

U.S. Fish and Wildlife Service, Region 1. "Bald Eagle Numbers Show Dramatic Growth; Nation's Symbol Eyed for Reclassification Under Endangered Species Act." News Release 90–03. February 7, 1990.

U.S. General Accounting Office. "Endangered Species Act: Types and Number of

Implementing Actions." Briefing Report to the Chairman, Committee on Science, Space and Technology, House of Representatives. Washington, D.C.: U.S. General Accounting Office, May 1992.

256

U.S. General Accounting Office. "Endangered Species: Management Improvements Could Enhance Recovery Program." Report to the Chairman, Subcommittee on Fisheries and Wildlife Conservation and the Environment, Committee on Merchant Marine and Fisheries, House of Representatives. Washington, D.C.: U.S. General Accounting Office, December 1988.

Chapter 8   The Rocky Return of the Gray Wolf

Associated Press. "When Bear No. 38 Wakes Up, What Will She Want to Eat?" *Idaho Falls Post Register*, March 1, 1985.

Bangs, Ed. "Return of a Predator: Wolf Recovery in Montana." *Western Wildlands*, Spring 1991.

Bath, Alistair J., and Phillips, Colette. *Statewide Surveys of Montana and Idaho Resident Attitudes Toward Wolf Reintroduction in Yellowstone National Park*. Report submitted to Friends of Animals, National Wildlife Federation, U.S. Fish and Wildlife Service and U.S. National Park Service. September 1990.

Cikaitoga, Tamra. "Bye-bye, Bear No. 38." *Idaho Falls Post Register*, September 4, 1984.

Environmental Defense Fund. Letter of September 19, 1990, to Manual Lujan, secretary, Department of the Interior.

Fischer, Hank, and Baden, John. "Carrots for Wolves." *Forest Watch*, July 1992, 10–11.

Freemuth, John. *Public Opinion on Wolves in Idaho: Results from 1992 Idaho Policy Survey*. Boise, Idaho: Department of Political Science, Boise State University, 1992.

Fritts, Steven H. "Gray Wolf Recovery in the Northern Rockies and Pacific Northwest." Helena, Mont.: U.S. Fish and Wildlife Service, February 1, 1992.

Hammer, Keith. "The Fate of a Female Wolf and Her Two Packs in Montana." *Predator Project Newsletter*, Fall 1992.

Mader, T. R. *Wolf Reintroduction in the Yellowstone National Park: A Historical Perspective*. Gillette, Wyo.: Common Man Institute, 1988.

Medberry, Mike. "Wolves in Idaho." *Forest Watch*, December 1991.

Tilt, Whitney; Norris, Ruth; and Eno, Amos. *Wolf Recovery in the Northern Rocky Mountains*. New York: Audubon Society/National Fish and Wildlife Foundation, April 1987.

U.S. Department of the Interior, National Park Service. "Re: Possible Wolf Sighted in Yellowstone National Park." News release. July 7, 1992.

U.S. Department of the Interior. *Wolves for Yellowstone?* National Park Service, Yellowstone National Park. Report to the U.S. Congress, Vol. III, Executive Summary. Denver, Colo.: National Park Service, July 1992.

U.S. Fish and Wildlife Service. *Wolf Recovery in Montana: First Annual Report, 1989*. Helena, Mont.: U.S. Fish and Wildlife Service, Fish and Wildlife Enhancement, 1989.

U.S. Fish and Wildlife Service. *A Summary of the Northern Rocky Mountain Wolf Recovery Plan.* Denver: U.S. Fish and Wildlife Service, 1987.

Weaver, John. "The Wolves of Yellowstone." Natural Resources Report No. 14. U.S. Department of the Interior, National Park Service. Washington, D.C.: U.S. Government Printing Office, 1978.

Wise, Carla; Yeo, Jeffrey J.; Goble, Dale; Peek, James M.; and O'Laughlin, Jay. "Wolf Recovery in Central Idaho: Alternative Strategies and Impacts." Report No. 4. Moscow, Idaho: Idaho Forest, Wildlife and Range Policy Analysis Group, University of Idaho, February 1991.

Chapter 9 Protecting the Wild Rockies

Bader, Mike. "The Northern Rockies Ecosystem Protection Act: A Citizen Plan For Wildlands Management." *Western Wildlands,* Summer 1991.

Craighead, John. "Yellowstone In Transition." Chapter 3 in *The Greater Yellowstone Ecosystem: Redefining America's Wilderness Heritage.* New Haven: Yale University Press, 1991.

Cutler, M. Rupert. "Needed: An American Biodiversity Network." *Defenders* Special Report, undated.

Farney, Dennis. "Unkindest Cut." *Wall Street Journal,* June 18, 1990.

Foreman, Dave, and Wolke, Howie. "The Big Outside: A Descriptive Inventory of the Big Wilderness Areas of the U.S. Tucson: Ned Ludd Books, 1989.

Kenworthy, Tom. "Unraveling of Ecosystem Looms in Oregon Forests." *Washington Post,* May 15, 1992.

Manning, Richard. "Timber Liquidation Was a Boardroom Decision." *Missoulian,* October 16, 1988.

Manning, Richard. "How a Montana Reporter Wrote What He Saw and Lost His Job." *High Country News,* September 23, 1991.

Picton, Harold D. "A Possible Link Between Yellowstone and Glacier Grizzly Bear Populations." *Bears: Their Biology and Management,* 1990.

Porterfield, Andrew. "Railroaded: The LBO Trend on Wall Street Is Playing Havoc with the Nation's Forests." *Common Cause,* September/October 1989.

Power, Thomas M. "The Timber Employment Impact of the Northern Rockies Ecosystem Protection Act in Idaho." Unpublished report. June 1992.

Power, Thomas M. "The Employment Impact of the Northern Rockies Ecosystem Protection Act in Montana." *The Networker,* Spring 1992.

Rasker, Ray; Tirrell, Norma; and Koepfer, Deanne. *The Wealth of Nature.* Washington, D.C.: The Wilderness Society, January 1992.

Richert, Kevin. "Montana Hesitant to Return Caribou." *Idaho Falls Post Register,* December 30, 1990.

Richert, Kevin, and Cecil, Molly O'Leary. "Threatened Habitats Imperil Threatened Species." *Idaho Falls Post Register,* December 30, 1990.

Rubovits, Robert. "Plum Creek: A Report on the Company's Environmental Policies and Practices." New York: Council on Economic Priorities, May 1992.

Scott, Kenneth. "Champion: A Report on the Company's Environmental Policies and Practices." New York: Council on Economic Priorities, May 1992.

Scott, Michael. "The Woodland Caribou." In *Audubon Wildlife Report*. New York: National Audubon Society, 1985.

258 Swanson, Cindy Sorg; McCollum, Daniel W.; and Maj, Mary. "Insights into the Economic Value of Grizzly Bears in the Yellowstone Recovery Zone." Draft Paper for the Forest Service Branch of the U.S. Department of Agriculture, 1992.

Chapter 10   Saving All the Parts

Baden, John A. "A Draft Paper." Seattle: Foundation for Research on Economics and the Environment, May 22, 1992.

Barker, Rocky. "Hedging the Bets." *Idaho Falls Post Register*, December 31, 1990.

DeBonis, Jeff. "Retrenchment in the Forest Service: Hardliners Oust Mumma in the Northern Rockies." *Inner Voice*, Fall 1991, 2–4.

Drushka, Ken. "The New Forestry: A Middle Ground in the Debate Over the Region's Forests?" *The New Pacific*, Fall 1990, 7.

Echard, Jo Kwong. *Protecting the Environment: Old Rhetoric, New Imperatives*. Washington, D.C.: Capitol Research Center, 1990.

Egan, Timothy. "Forest Supervisors Say Politicians Are Asking Them to Cut Too Much." *New York Times*, September 16, 1991.

Fischer, Hank, and Baden, John. "Carrots for Wolves." *Forest Watch*, July 1992, 10–11.

Greater Yellowstone Coordinating Committee. "Vision for the Future: A Framework for Coordination in the Greater Yellowstone Area." Draft copy. Denver, Colo.: U.S. National Park Service and U.S. Forest Service, August 1990.

Groves, Craig. "Beyond Endangered Species, Gap Analysis." *Idaho Wildlife*, Winter 1992, 26–28.

Kellert, Stephen R., and Clark, Tim W. "A Framework for Wildlife Policy Understanding and Analysis." *Policy Studies Journal*. In press.

Maser, Chris, and Franklin, Jerry. "From the Forest to the Sea: A Story of Fallen Trees." U.S. Department of Agriculture General Technical Report 229. Portland: USDA, Forest Service, 1988.

O'Toole, Randal. "Appropriate Economics, Ethics and Evolutionarily Stable Systems." *Forest Watch*, May 1992.

Reese, Rick. *Greater Yellowstone: The National Park and Adjacent Wildlands*. Helena, Mont.: American and World Geographic Publishing, 1991.

Richert, Kevin. "Vision Report Symbolizes Problems of Protecting an Ecosystem." *Idaho Falls Post Register*, March 18, 1991.

Robertson, Dale. *Ecosystem Management of the National Forests and Grasslands*. Washington, D.C.: USDA, Forest Services, June 4, 1992.

U.S. Environmental Protection Agency. "Reducing Risk: Setting Priorities and Strategies for Environmental Protection." Washington, D.C.: U.S. Environmental Protection Agency, September 1990.

Index

About the Author

Roland "Rocky" Barker is the senior reporter in charge of special projects for the *Post Register* in Idaho Falls, Idaho. In 1990, he led the newspaper's team of reporters, editors and photographers in presenting an exhaustive study of the Endangered Species Act and its effects on the Pacific Northwest and northern Rockies. The series was a runner-up for the Edward J. Meeman Award, the top environmental journalism award in the country, and the winner of numerous state and regional awards. Barker, forty, also was the lead reporter for the *Post Register*'s award-winning coverage of the 1988 Yellowstone fires and its nationally recognized coverage of nuclear waste problems at the Idaho National Engineering Laboratory.

He holds a bachelor's degree in environmental studies from Northland College in Ashland, Wisconsin, and is a fellow of the Knight School of Specialized Journalism at the University of Maryland. Barker is a regular contributor to *High Country News* and has been published in the *Los Angeles Times, Chicago Tribune, Milwaukee Journal, Minneapolis Tribune, Wisconsin Sportsman, Wisconsin Natural Resources, Fin and Feathers, Buzzworm* and *Outside*.

Barker and his wife, Tina, have two fourteen-year-old sons and a seven-year-old daughter and live in Idaho Falls.

Island Press Board of Directors

SUSAN E. SECHLER, CHAIR, Director, Rural Economic Policy Program, Aspen Institute for Humanistic Studies

HENRY REATH, VICE-CHAIR, President, Collector's Reprints, Inc.

DRUMMOND PIKE, SECRETARY, President, The Tides Foundation

ROBERT E. BAENSCH, Senior Vice President/Marketing, Rizzoli International Publications, Inc.

PETER R. BORRELLI, Executive Vice President, Open Space Institute

CATHERINE M. CONOVER

PAUL HAWKEN, Chief Executive Officer, Smith & Hawken

CHARLES C. SAVITT, President, Center for Resource Economics/ Island Press

PETER R. STEIN, Managing Partner, Lyme Timber Company

RICHARD TRUDELL, Executive Director, American Indian Resources Institute